CREATING FRESHWATER WETLANDS

DONALD A. HAMMER

LEWIS PUBLISHERS
Boca Raton   Ann Arbor   London

Library of Congress Cataloging-in-Publication Data

Hammer, Donald A.
Creating freshwater wetlands / by Donald A. Hammer
    p. cm.
  Includes bibliographical references and index.
  ISBN 0-87371-445-8
  1. Construction and maintenance. 2. Hydrology. I. Title
TD607.E7H564   1992
633.2′0075′8—dc20   92-9833
                                                          CIP

LEWIS PUBLISHERS
121 South Main Street, Chelsea, MI 48118

PRINTED IN THE UNITED STATES OF AMERICA
  3 4 5 6 7 8 9 0

PREFACE

Brian Lewis conceived of this little handbook as a supplement to the 1988 conference proceedings *Constructed Wetlands for Wastewater Treatment*, and at his urging, I finally committed to begin compiling the information.

As a student I was fortunate in working on several National Wildlife Refuges, and in one case, my research benefited from the historical information on wildlife populations in old annual reports. However, in digging out information on snapping turtles from the earliest days of the refuge, I discovered the tremendous wealth of information on methods developed to restore and create the marsh that I believed was largely natural. Later I dicovered similar knowledge in the earliest reports from virtually every refuge and management area that I visited.

Since many of our "natural" wetlands today were in fact restored or even created, I have been surprised when colleagues state that we know so little about wetlands that we can't restore or create one as mitigation or for any other reason. In fact, the predominant conclusion from most recent wetlands meetings would be our total ignorance of basic wetlands ecology. Certainly we don't know everything about wetlands — they might not be as fascinating to some of us if we did — but to conclude that we lack the information to restore or create many types of wetlands is to ignore or negate a considerable body of information gathered by competent, dedicated scientists and managers over many years.

Doubtless some will respond that this information only applies to marshes. Certainly much of it does, but some is also available for other types, because bogs, swamps, sloughs and fens were included in wildlife refuges and management areas. But more importantly, marshes are precursors, early successional stages for many other wetland ecosystems. If we can create a healthy marsh and then allow it to progress along the successional gradient to another wetlands type, are we not well on the way to creating the bog or swamp that is the project goal? And do we know how to create a marsh and allow it to transition? Of course we do. We simply do not supply the disturbance factor that is critical to maintaining the site in the early marsh stage.

Although today's society places considerable importance on natural and created wetlands, it was not always so. But while most of past society had little use or regard for "dark and dismal swamps," and wetlands drainage was national policy, a few conservationists began raising the alarm over lost wetlands as early as the turn of the century. Extensive drainage projects coupled with the severe drought of the 1930s caused calamitous declines in North American waterfowl populations. Alarmed duck hunters, many wealthy and influential in society, initiated legislation taxing sport hunters to support wetlands acquisition, management and research that provided the basis for the National Wildlife Refuge System and comparable state wildlife management areas. Faced with a dirth of water and wetlands habitats but charged with restoring waterfowl populations, early managers slowly developed methods to create and restore natural wetland ecosystems to enhance waterfowl habitats throughout the country.

Creation of various public works programs (CCC, WPA, etc.) concurrent with the drive to restore waterfowl provided heretofore unheard of funding and manpower resources in the newly emerging field of wildlife management. Substantial efforts in these programs were directed towards developing methods to design and construct dikes and water control structures, plant and/or seed wetlands and terrestrial vegetation, re-stock wildlife and fish populations, manipulate water levels and manage fish and wildlife populations in the restored or created wetlands ecosystems. Probably because of the bureaucratic penchant for documentation, most of the methods developed and tested were reported in the "gray literature" — the annual reports for each refuge or management area. Not surprisingly, virtually all types of wetlands ecosystems in the U. S. are encompassed. Simple examination of a distribution map for the National Wildlife Refuges reveals refuge locations from coast to coast, including most types of freshwater wetlands and some brackish/saltwater systems. Restoration and creation techniques were investigated and reported in the late 1930s, 1940s and 1950s on most of these areas.

And early managers unrestrained by the policy of not reporting "negative results" prevalent in scientific journals today, included failures as well as successes. Unfortunately much of this knowledge never found its way into scientific journals and, hence, is largely unknown to the new crop of wetlands ecologists and managers. Some has not even been written down though it has been passed along and is still in use by present managers. But that information is still available in the files or from the staff of the local refuge or management area office.

Unfortunately many sincere and dedicated wetlands enthusiasts are unaware of this wealth of information, or choose to disregard it — "they were only interested in growing ducks." I am afraid I must disagree with the latter. Techniques developed in the late 1930s and 1940s to establish or manage *Scirpus, Potamogeton, Spartina, Quercus, Larix* or *Taxodium* or a host of other wetlands plants need not be reinvented today, nor must we use the most expensive methods. Though propagation and hand-planting were used, early wetlands managers slowly discovered the growth requirements for important species and developed methods to foster or retard vegetative types through inexpensively manipulating water levels in appropriate seasons.

Methods developed for management of duck production, migratory or wintering areas have successfully provided wetlands habitats for innumerable other species of wetlands wildlife, fish and plants. Ducks may have been the priority but "they didn't just grow ducks." In fact, many of the larger "natural" wetlands extant today are located in refuges, management areas and a few parks because they were restored, enhanced and in a few instances, created by early waterfowl managers.

Finally, my earlier experiences creating or managing wetlands with the life support function as a primary goal and more recent work with the water purification function have convinced me that a wetlands ecosystem which supports diverse and abundant vertebrate populations does so because it has the complex, requisite structure. If the structure is present to provide significant life support, then that system likely has the structure to provide much of the water purification, hydrologic buffering and other important functions that we value in wetland ecosystems. Regardless of the prime objective, designing and constructing a wetlands that will attract and support representative vertebrate species will likely achieve the prime objective plus support many other important functional values.

This then is a small attempt to organize and present some of the information on methods to create or restore freshwater wetlands accumulated by wetland scientists and managers during the last 50 years. It is my hope that the reader will interpret this volume as a general guideline and employ specific, localized information from nearby offices of wetlands and waterfowl research/management agencies before starting a project.

D. A. Hammer
Norris, Tennessee
2 March 1991

TABLE OF CONTENTS

1   MARSHES, BOGS, SWAMPS, SLOUGHS, FENS,TULES, AND BAYOUS

INTRODUCTION

Today, wetlands are on everyone's tongue, but just 10 years ago only a few biologists knew of the term and fewer wetlands specialists partially understood these complex systems. Now, even the U.S. President proclaims his dedication to preserving these important resources and a fashionable restaurant on Hudson Street in downtown Manhattan is named "Wetlands." In fact, the term has only recently come into common usage to provide a generic, all encompassing word that includes virtually all types of shallow water environments.

In the past, a few sportsmen, conservationists, and scientists were interested in managing and protecting marshes, swamps and bogs — wetlands. The vast majority of society viewed swamps and bogs as obstacles to progress since most wetlands were thought to be reservoirs of disease and unfit for farming or development, if not actually haunts of unimaginable monsters. In 1977, President Carter issued Executive Order 11990 — Protection of Wetlands — and 13 years later, the change in society's attitudes has been dramatic. The change reflects the accomplishments of a few dedicated conservationists that managed to bring the issue to the public's mind, but the surprising speed of the shift suggests a significant proportion of society was sympathetic. However, few were likely to have been simply supportive of wetlands — most of society had little or no concept of wetlands or their values to society. Most likely, the wetlands issue was assimilated into the overall concern for world environments during a period of elevated human consciousness.

Few lay supporters have more understanding of wetlands resources, wetlands ecology, or real functional values of wetlands than they do for acid rain, the ozone hole, or rain forests — all current issues with fervent, vocal proponents having substantial public concern. Many wetlands protectors could not define wetlands, much less describe, the complex biological communities and processes within the wide range of wetlands types. Many that would use wetlands today have little understanding for the complex systems they attempt to replicate or reduce for their own purposes. Even among wetlands scientists, there remains much disagreement on precise definitions and limited understanding of biological communities and hydrologic, physical, and chemical processes.

Our ideas of wetlands are vague and ambiguous, partially because of the bewildering variety and broad range of environments encompassed within the term. Some types of wetlands are commonly recognized and the names widely known; but names vary confusingly in different regions and are often encumbered with historical human perceptions. For example, a "slough" is a freshwater marsh in the Dakotas, a brackish marsh along the West Coast, and a freshwater swamp on an old river channel in the Gulf Coastal Plain. Marsh has had limited use, but slough, swamp, and bog in common usage have become burdened with foreboding, difficult, or hindering connotations. Since widespread interest is only recent, limited information is available, even in scientific fields; and definitions are important for regulatory actions, considerable research and discussion has occurred during the last 15 years. Since land and water can mix in many ways and biotic components modify and blur the boundaries, it can be perplexing to define wetlands or determine where wetlands begin or end strictly on wetness or dryness. Not surprisingly, definitions, especially on precise wetlands boundaries, are difficult to derive and apply. That is simply the indefinite nature and complexity of the subject, not a reflection on the ability of those studying it.

Much of the confusion derives from the indefinite character of wetlands ecosystems. Wetlands are ecotones (edges), transition zones between dry land and deep water, environments that are not always wet nor obviously dry. Any sizable wetlands often include portions that are clearly dry land and clearly deep water, as would be expected in any transition region. Boundaries are imprecise and may seem to vary with seasons and different years. Gradual changes in wetness, soil, and vegetation types occurring across the transition band confound attempts to precisely measure boundaries and subsequently process descriptions. Furthermore, wetlands need not be continuously wet, nor are they continuously dry. Many wetlands are only wet during certain years, seasons, times of day, or after heavy rains. At other times, they may be dry. However, the unique plant and animal communities in wetlands ecosystems depend on environmental conditions created by alternating inundation and drying during different seasons or different years.

The rich variety of plants and animals found in most wetlands results from their transitional position in the landscape and subsequent production rates (see Figure 1). Not only are many unique organisms restricted to wetlands environments, but most wetlands receive extensive use by animals characteristic of terrestrial or purely aquatic environments. Some use wetlands seasonally — various fish spawn in shallow water wetlands, but spend most of their adult lives in deeper waters. Others visit daily — fox or coyotes on their nightly rounds; while others may reside for extended periods, depending on availability of other foraging and shelter conditions — deer or pheasants weathering winter storms or antelope and elk browsing succulent marsh vegetation. Many birds found in terrestrial and wetlands habitats frequently have their highest numbers in the diverse, productive habitats of wetlands.

Diversity and abundance vary greatly between different types of wetlands and within a single wetlands. Some wetlands — acidic bogs, monotypic cattail (*Typha*), or reed (*Phragmites*) marshes and many saltwater wetlands — have low diversities (i.e., large numbers of a few types of plants or animals). Others, river swamps and fresh/ brackish marshes, have high diversities; that is, many types of plants or animals, but only a few individuals of each type. In either case, basic productivities measured as biomass produced per unit area per unit time commonly exceed the production rates for the most intensively managed agricultural fields.

Figure 1.   The rich variety and high productivity of this prairie marsh results from its transitional location between dry upland and deep water environments.

Variation in productivity and diversity within a wetlands system is readily apparent from casual observation of the "hummocks" within the Everglades. These wet, forested islands situated in large expanses of wet prairie, "the river of grass," support a more diverse assemblage of plant and animal species than the adjacent sedge marshes and mangrove swamps. Hummocks also provide critical seasonal habitats for animals normally found in the marshes when the latter become too wet or too dry for certain species. Consequently, the diversity and productivity of hummocks varies substantially during the course of a year, and their influence extends far beyond their boundaries.

WETLAND DEFINITIONS

Problems in defining wetlands for all uses is reflected in the variety and types of earlier definitions, many of which were devised for different needs and purposes. Most avoid the how-wet-is-wet question by describing wetlands in terms of soil characteristics and the types of plants capable of growing in these wet transitional habitats. Even shallow standing water or saturated soil quickly causes the atmospheric gases that filled interstitial pore spaces in the soil to be replaced by water, and microbial metabolism rapidly consumes available oxygen. Since gaseous diffusion from the atmosphere into soil water is much slower than microbial consumption, all except a thin top layer of the soil becomes anoxic or without oxygen.

Roots of normal, terrestrial plants obtain oxygen for respiration from gases within soil pore spaces and if those spaces are filled with water lacking oxygen, their roots die and the plant dies. Hydrophytic or wetlands plants have developed specialized physical structures, aerenchyma, loosely similar to bundles of drinking straws, to transport atmospheric gases including oxygen through leaves and stems down to the roots to provide oxygen for respiration. Aerenchyma also transport respiratory by-products and other gases generated in the substrate back up the roots, stem, and leaves for release to the atmosphere, reducing potentially toxic accumulations in the region of growing roots. Because of these specialized structures, wetlands plants are able to survive and grow in habitats with hostile root-growing conditions that would kill other plants. Consequently, wetlands plants are often the best indicator of a wetlands system even though many wetlands plants can grow in drier environments if competition with terrestrial plants is limited.

Inundation and anaerobic conditions also cause specific changes in chemical substances found in most soils that serve as indicators of wet soils. Anoxic substrates with reducing environments cause many elements and compounds to occur in reduced states, creating characteristic colors, textures, and compositions typical of hydric soils. Due to the prevalence of iron in many soils and its color in reduced states, wet soils often have a gray or grayish color and fine texture.

In 1979, the U.S. Fish and Wildlife Service developed a generic definition and classification system to encompass and systematically organize all types of wetland habitats for scientific purposes. It broadly recognizes wetlands as a transition between terrestrial and aquatic systems, where water is the dominant factor determining development of soils and associated biological communities and where, at least periodically, the water table is at or near the surface, or the land is covered by shallow water. Specifically, "Wetlands must have one or more of the following three attributes:

1. at least periodically, the land supports predominantly hydrophytes;
2. the substrate is predominantly undrained hydric soil, and;
3. the substrate is nonsoil and is saturated with water or covered by shallow water at some time during the growing season of each year.

This definition broadens the three essential components of wetlands in the definition contained in the wetlands protection Executive Order. It concentrates on areas containing undrained or poorly drained (hydric) soils or areas with nonsoil substrates (rock or gravel) that are covered by water during a portion of the growing season. In either instance, continued inundation (soil or rock) precludes establishment or long-term survival of plants lacking special adaptations to growing in flooded substrates. Basically, these areas are wet enough for long enough to produce anaerobic substrate conditions that limit the types of plants that can survive there. Only wetlands plants (hydrophytes) with the ability to provide oxygen for root respiration from atmospheric sources will be present.

But importantly, two of three attributes include the qualification "predominantly" since only a rare wetlands would be completely lacking in at least a few small areas of normal aerobic soil or other substrates supporting typical terrestrial plants. Since few wetlands have perfectly flat surfaces or uniformly consistent elevation changes, most have portions that are essentially terrestrial habitats.

Conditions for wetlands soils and vegetation are produced by the impact of water, and extent and duration of flooding may vary substantially in some areas with only a few centimeters difference in elevations. In addition, even a perfectly flat swamp will support some terrestrial vegetation on living tree trunks and most certainly on the remains of stumps and fallen boles that extend slightly above the typical flood line. Muskrat houses, ice heaves, and herbivore wallows form similar high spots in prairie and coastal marshes. It is important to bear in mind that most (majority, dominant, etc.) of the area must consist of hydric soils and hydrophytic vegetation, but not necessarily 100% or even 80%. Since wetlands are transition zones, a mixing or merging of environmental conditions is expected and in fact an important characteristic that contributes to the diversity and productivity of our wetlands resources.

Also significant is the concept and implication of "periodically" and "at some time" in two attributes of this and other definitions. Both encompass alternating wet and dry periods — not necessarily continuously wet, but not continuously dry — that are critical in determining the types of vegetation that can survive there. Bottomland hardwoods (swamps) frequently sustain deep and often long-term inundation during winter, but similar conditions in spring and early summer cause physiological stress leading to death if flooding occurs over more than one growing season. Absence of winter flooding would remove the competitive advantage of wetlands trees vs. upland trees and also reduce production (biomass) since the annual fertilization and watering phenomenon would be lacking. Conversely, continuously flooding in a prairie marsh (over 5 to 10 years) causes falling productivity, eventually plant stress, and finally mortality leaving deeper, open water environments of shallow lakes. Alternating periodic flooding and drying are crucial to maintaining the complexity, diversity, and productivities of natural wetlands, but the concept and its manifestations in soils, hydrology, and biological communities confound attempts to develop simple definitions and precise boundary determinations.

NEW WETLAND DEFINITIONS

In 1990, a joint definition for wetlands developed and formally adopted by the Fish and Wildlife Service, the Environmental Protection Agency, the Soil Conservation Service, and the Corps of Engineers provides a common definition for federal agencies. Due to the pervasive nature of regulations promulgated by these agencies and their direct impacts on wetlands resources, this definition is expected to gain acceptance by state and local governments and other organizations. Though its language is only slightly different, it has been interpreted in a manner that places more emphasis on hydric soils and lessens the importance of wetlands vegetation in wetlands determinations. By extension, the historical conditions at any specific site become important since hydric soils are formed over fairly long time periods and at least some characteristics often persist for long periods even after adequate drainage is established.

Consequently, regulators are faced with explaining their description of a soy bean field or other cropland as wetlands even though little or no wetlands vegetation is present. This is almost "once a wetlands always a wetlands" even though truly hydric soils and wetlands vegetation may not have been present for tens of years. However, blocking drain lines or ditches would likely lead to gradual restoration of the historic wetlands ecosystem, providing a basis for defining the present-day crop field as a wetlands. New legislation and regulations embodying this concept penalize landowners for subsequent drainage and conversion to agriculture if a field has been abandoned for a prescribed time period. However, abandonment is defined in temporal terms and little or no consideration is given to whether or not hydric soils and wetlands vegetation have returned or other components of natural wetlands form and function have or are likely to be reestablished.

Unfortunately, unless limited efforts at present are expanded, many acres of potential wetlands or cropfields may become low-value hybrids. Restoration of the nation's wetlands is important and valuable to our society but expecting an individual landowner to forego crop production on his land is expecting inordinate contributions from a small minority of society. Neither will the reverting croplands provide significant functional values expected of natural wetlands in any reasonable time interval. Society must develop effective means of compensating the landowner for lost acreages while simultaneously providing him with the methods and perhaps financial means to initiate active restoration efforts. Since many areas of converted wetlands are marginally productive agricultural lands, expansion of the current efforts to prioritize locations, quantities, and qualities of national wetlands resources are urgently needed.

Once a general consensus for high-priority areas has been reached, leasing, easements, or acquisition and active restoration efforts should be initiated.

Although a wetlands definition was only recently standardized, efforts to protect, restore, create, or use wetlands for specific purposes have already required additional modifiers. Foremost among these efforts has been creating wetlands for mitigation purposes and close behind are projects building wetlands for water purification. In each instance, a need arose to clearly distinguish between wetlands built for specific functions, especially water treatment, those built to mitigate wetlands impacted by development, and natural wetlands for regulatory purposes. Since revisions to the Clean Water Act, specifically Section 404, have made it the principal regulatory tool for protection of natural wetlands, the Portland Regional Office of EPA developed a set of definitions and interpretations to differentiate between natural wetlands and man-made systems for application in the 404 permit review process.

> "Constructed wetland: Those wetlands intentionally created from nonwetland sites for the sole purpose of wastewater or stormwater treatment. These are not normally considered waters of the U.S.
> Constructed wetlands are to be considered treatment systems (i.e., not waters of the U.S.); these systems must be managed and monitored. Upon abandonment, these systems may revert to waters of the U.S. Discharges to constructed wetlands are not regulated under the Clean Water Act. Discharges from constructed wetlands to waters of the U.S. (including natural wetlands) must meet applicable NPDES permit effluent limits and state water quality standards.
> Created Wetland: Those wetlands intentionally created from nonwetland sites to produce or replace natural habitat (e.g., compensatory mitigation projects). These are normally considered waters of the U.S.
> Created wetlands must be carefully planned, designed, constructed, and moni-tored. Plans should be reviewed and approved by appropriate state and federal agencies with jurisdiction. Plans should include clear goal statements, proposed construction methods, standards for success, a monitoring program and a contin-gency plan in the event success is not achieved within the specified time frame. Created wetlands should be located where the 'return' to the environment will be maximized (not necessarily onsite) and should be protected in perpetuity, to the extent feasible, through easements, deed restrictions, or transfer of title to an appropriate conservation agency or organization. Site characteristics should be carefully studied, particularly hydrology and soils, during the design phase and created wetlands should not be designed to provide habitat and provide stormwater treatment."

Natural wetlands were not newly defined, but as in the above, guidelines and restrictions on use were emphasized.

> "Discharges to a natural wetland must not degrade the functions/beneficial uses of the wetland (i.e., must meet state water quality standards applicable to the wetland and comply with EPA and state antidegradation policies). All practicable source control best management practices must be applied to minimize pollutants entering the wetland, consistent with NPDES permit requirements. Source control BMPs would generally include erosion controls, oil/water separation, presettling basins, biofilters, etc. Inlet/outlet structures requiring fill must be permitted under Section 404 of the Clean Water Act, preferably via an individual permit. Natural

wetlands may not be used for instream treatment in lieu of source controls/ advanced treatment; may be used for 'tertiary' treatment or 'polishing' following appropriate source control and/or treatment in a constructed wetland, consistent with the preceeding guidelines."

The emphasis on careful planning, construction, contingency plans, etc. for created wetlands, most of which are built for mitigation purposes, reflects a regulatory perspective and a growing pessimism over the ability of designers and developers to successfully replace lost or damaged natural wetlands. Many have failed. Some because of overly ambitious plans, limited time for natural successional stages or poor designs and construction. However, evaluation is often difficult since objective, quantitative goals were rarely included in original plans. Obviously, the created wetlands definition and interpretations are designed to establish a new wetlands that will replace one lost through development. However, only two possible uses are discussed though many others could be incorporated and additional guidance should specify management, as necessary, to ensure that the new wetlands will not only develop the form (structure in terms of water, soil, and biological communities) but also the functions performed by the replaced wetlands that are valued by society.

The perspective embodied in the created wetlands discussion derives from the regulatory need to clearly distinguish mitigation wetlands from wetlands wastewater treatment systems and to ensure the success of wetlands mitigation projects; that is, those proposed as replacements for damaged or destroyed wetlands. However, as written, it would as easily apply to wetlands built for wildlife habitats, hydrologic buffering, or recreational purposes. In most cases, the latter would be considered waters of the U.S. and subject to the provisions of Section 404. In the current regulatory climate, fear of potential regulatory complications is likely to discourage landowners interested in building a wetlands for recreation or other nontreatment functions. Furthermore, most newly built wetlands will require considerable active management before becoming fully established and many will require periodic management to maintain a specific successional stage. Since maintenance might include water level manipulation, even total drying, controlled burning, or other disturbances, these activities would seem to be regulated and perhaps would be prohibited under this interpretation. Quite obviously, we wish to protect and preserve wetlands built as replacements but just as clearly, we must not discourage increasingly widespread and significant efforts to build wetlands in nonwetland sites as a means of increasing our total wetlands resource base. Consequently, created wetlands should also include those wetlands built for nonmitigation purposes, but should exempt the latter from current and future inclusion in the waters of the U.S.

It is also clear that deliberate use of natural wetlands for water treatment purposes is unacceptable except for advanced or polishing treatment. Though we understand how to build and operate a constructed wetlands for efficient water treatment, most are simply that, wastewater treatment plants and the above discussion properly recognizes their status and emphasizes needs to manage and monitor constructed wetlands treatment systems. At present, our limited understanding of water purification mechanisms and processes within constructed wetlands treatment systems is inadequate to recommend application rates that would not impair other functional values of natural wetlands. Until this new technology has progressed considerably, natural wetlands should be protected from deliberate applications of sediments and anthropogenic pollutants. In addition, many natural wetlands are presently receiving moderate to high loadings of pollutants generated by point and nonpoint sources and the impacts of these, much less additional deliberate applications, are unknown.

WETLANDS CLASSIFICATION

The wetlands *classification* system is a hierarchical system similar to those used for classifying plants and animals. It starts with 5 large systems; these are progressively divided into 10 subsystems, 55 classes, and 121 subclasses, which are then characterized by examples of dominant types of plants or animals. This system provides a consistent standard of terminology for use among scientists and managers throughout the country. However, careful and consistent determination of presence or absence of wetlands from *definitions* must be employed prior to use of the classification systems since the latter is capable of *classifying* almost any area that is periodically wet, even a rain puddle in a parking lot. The classification system is a tool to categorize or *classify* a wetlands following the application of some other appropriate *definition* to determine whether the area in question is a wetlands. It should not be used to determine whether or not a site has a wetlands on it.

The classification system is commonly used by wetlands specialists and it provides the framework for the National Wetlands Inventory, a comprehensive identification and mapping of wetlands led by the U.S. Fish and Wildlife Service. The standardized procedures for delineation and quantification will provide a resource database for evaluating quality and quantity of remaining wetlands and assessing negative and positive impacts of destructive as well as creation or enhancement developments.

However, nonspecialists and various legislation and agency regulations will continue to define wetlands in more general terms. With minor variations, most describe wetlands as areas flooded or saturated by surface water or groundwater often and long enough to support those types of vegetation and aquatic life that require or are specially adapted for saturated soil conditions. Such descriptions can accommodate much of the conceptual framework and detailed, specific terminology necessary for scientific classifications. Concurrently, the generic terms more closely adapt to popular conceptions of what constitutes wetlands — salt and freshwater swamps, marshes, and bogs and perhaps a few subcategories of these basic types.

In popular usage, shallow-water or saturated areas dominated by water-tolerant woody plants and trees are generally considered swamps; those dominated by soft-stemmed plants such as cattail and bulrush are considered marshes, and those with mosses and evergreen shrubs are bogs.

Our principal saltwater swamps are mangrove wetlands along the southern coast of Florida. Mangroves are among the very few woody plants adapted to saltwater environments (see Figure 2).

Coastal salt marshes (see Figure 3) are dominated by salt-tolerant herbaceous plants, notably cordgrass (*Spartina*), blackrush (*Juncus*) or other rushes along exten-

Figure 2.   Drought exposes the prop roots of newly established mangroves that will accumulate sediments and organic materials to form islands in the Everglades region of Florida.

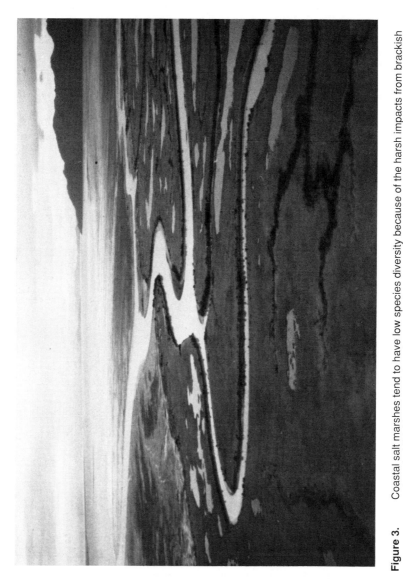

Figure 3.   Coastal salt marshes tend to have low species diversity because of the harsh impacts from brackish waters, but tidal transport of nutrients contributes to high productivities.

Figure 4.   Cypress and a few other tolerant trees can withstand almost permanent inundation, but most hardwood swamps require drying during the growing season.

sive areas of the eastern and southern coasts, or cordgrass and glasswort (*Salicornia*) along the west coast. Less familiar are the inland salt marshes of the intermountain west where high evaporation rates from shallow lakes and playas concentrates salt contents favoring similar plant types.

Freshwater swamps contain a variety of woody plants and water-tolerant trees. Southern swamps typically contain bald cypress (*Taxodium*), tupelo gum (*Nyssa*), water, willow oak, and swamp white oak (*Quercus*), and river birch (*Populus*). Northern swamps are more likely to include alder (*Alnus*), black ash (*Fraxinus*), black gum (*Nyssa*), northern white cedar (*Thuja*), black spruce (*Picea*), tamarack (*Larix*), red maple (*Acer*), and willow (*Salix*). Forested wetlands include bay swamps, peat swamps, white cedar swamps, wet flats, muck swamps, cypress heads or strands, bottomland hardwoods, and mesic riverine forests (see Figure 4).

Freshwater marshes are dominated by herbaceous plants. Submerged and floating plants may occur, often in abundance, but emergent plants usually distinguish a marsh from other aquatic environments. Familiar emergents include cattails (*Typha*), bulrush (*Scirpus*), reed (*Phragmites*), grasses, and sedges (*Carex*) (see Figure 5). A wet meadow may be only intermittently saturated or flooded with very shallow water, but it also supports marsh species, especially sedges and wet grasses. A common type of freshwater marsh, the prairie potholes of the northern Great Plains, occurs in shallow depressions formed by glaciers. Those that hold water year-round, seasonally, or following heavy rains, often support luxuriant marsh vegetation. Although most are small, there are many of them (810,000 ha in North Dakota alone), and collectively they constitute an important wetland resource, especially for waterfowl nesting.

Bogs form primarily in deeper glaciated depressions, mainly in "kettle holes" in the northeastern and northcentral regions (see Figure 6). Bogs are dependent upon stable water levels and are characterized by acidic, low-nutrient water and acid-tolerant mosses. Other bog plants such as cranberry (*Vaccinium*), tamarack, black spruce (*Picea*), leatherleaf (*Chamaedaphne*), and pitcher plant (*Sarracenia*) may be rooted in deep, spongy accumulations of dead *Sphagnum* moss and other plant materials only partially decomposed under bog conditions. In the same region and also at high elevations in the Rockies, fens have water nearer to neutral and are dominated by sedges.

FUNCTIONS OF NATURAL WETLANDS

Wetlands represent a very small fraction of our total land area, but they harbor an unusually large percentage of our wildlife. For example, 900 species of wildlife in the U.S. require wetland habitats at some stage in their life cycle, with an even greater number using wetlands periodically. Representatives from almost all avian groups use wetlands to some extent and one third of North American bird species rely directly on wetlands for some resource.

Due to the diversity of habitats possible in these transition environments, the nation's wetlands are estimated to contain 190 species of amphibians, 270 species of birds, and over 5000 species of plants. Many wetlands are identified as critical habitats

Figure 5.   Cattails, bulrush, and sedges are typically dominant species in the freshwater marshes of the prairie pothole region.

Figure 6.   Stable water levels with low nutrients and acidic conditions favor establishment of bogs in New England depressions.

under provisions of the Endangered Species Act, with 26% of the plants and 45% of the animals listed as threatened or endangered either directly or indirectly dependent on wetlands for survival.

Stability is neither common nor desirable in wetland systems. Unlike upland habitats, wetlands are dynamic, transitional, and dependent on natural perturbations. The most visible and significant perturbation is periodic inundation and drying. Changing water depths, either daily, seasonal, or annual, strongly influence plant species composition, structure, and distribution. Other influences, such as complex zones of water regimes, salt and temperature gradients, and tide and wave action, produce wetlands vegetation that is generally stratified, much like forests. These factors combine to create a diversity and wealth of niches that make wetlands important wildlife habitat.

In addition to their vegetative productivity, wetlands team with zooplankton, worms, insects, crustaceans, reptiles, amphibians, fish, birds (see Figure 7), and mammals, all feeding on plant materials or on one another. Other animals are drawn from nearby aquatic or terrestrial environments to feed on plants and animals at the highly productive "edge" environment of wetlands, and they in turn become prey for others from a greater distance, thus extending the productive influence of wetlands far beyond their borders.

Sport and commercial hunters and fishermen have called public attention to the economic value of wetlands fish and wildlife. They were first to note the direct relationship between wetland destruction and declining populations of valuable species of fish, shellfish, birds, reptiles, and fur-bearing animals that are dependent on certain types of wetland habitats during part or all of their lives. Many studies have now linked the destruction of summer breeding wetlands and winter feeding wetlands to shifts or declines in populations of migratory waterfowl and other birds. Wetland destruction can be especially significant in regions where such habitat is least common and alternative sites may be unavailable.

Wetlands along coasts, lakeshores, and riverbanks have recently begun receiving increased attention because of their valuable role in stabilizing shorelands and protecting them from the erosive battering of tides, waves, storms, and wind. One of the greatest benefits of inland wetlands is the natural flood control or buffering certain wetlands provide for downstream areas by slowing the flow of floodwater, desynchronizing peak contributions of tributary streams, and reducing peak flows on main rivers. During dryer periods, slow releases from wetlands augment and stabilize base flows providing water to support aquatic life in streams and rivers.

Some wetlands may function as groundwater recharge areas, allowing water to seep slowly into and replenish underlying aquifers. Other wetlands represent discharge areas for surfacing groundwaters. Both may occur within close proximity depending upon local and regional patterns of ground water distribution. This may be one of the reasons that simply plugging the ditch from a drained wetlands does not always restore a productive marsh. The previous system may have been the expression of surfacing ground water that previously was supplied by another still-drained wetlands that acted as a recharge area. Some of these interdependent wetlands may lie fairly close together, but others may be many kilometers apart.

Another important but poorly understood function is water quality improvement.

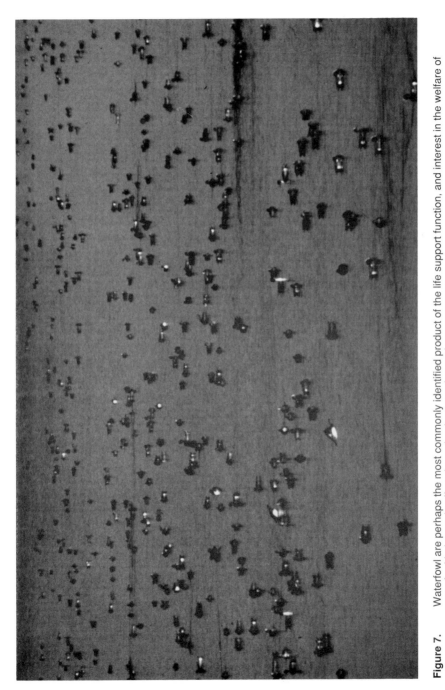

Figure 7.   Waterfowl are perhaps the most commonly identified product of the life support function, and interest in the welfare of waterfowl led to early efforts in wetlands protection and management.

Wetlands provide effective, free treatment for many types of water pollution. Wetlands can effectively remove or convert large quantities of pollutants from point sources (municipal and certain industrial wastewater effluents) and nonpoint sources (mine, agricultural, and urban runoff), including organic matter, suspended solids, metals, and excess nutrients. Natural filtration, sedimentation, and other processes help clear the water of many pollutants. Some are physically or chemically immobilized and remain there permanently unless disturbed. Chemical reactions and biological decomposition break down complex compounds into simpler substances. Through absorption and assimilation, wetland plants remove nutrients for biomass production. One abundant by-product of the plant growth process is oxygen, which increases the dissolved oxygen content of the water and also of the soil in the immediate vicinity of plant roots. This increases the capacity of the system for aerobic bacterial decomposition of pollutants as well as its capacity for supporting a wide range of oxygen-using aquatic organisms, some of which directly or indirectly utilize additional pollutants.

Many nutrients are held in the wetland system and recycled through successive seasons of plant growth, death, and decay. If water leaves the system through seepage to groundwater, filtration through soils, peat, or other substrates removes excess nutrients and other pollutants. If water leaves over the surface, nutrients trapped in substrate and plant tissues during the growing season do not contribute to noxious algae blooms and excessive aquatic weed growths in downstream rivers and lakes. Excess nutrients from decaying plant tissues released during the nongrowing season have less effect on downstream waters.

It is no secret that natural wetlands can remove iron, manganese, and other metals from acid drainage — they have been doing it for geological ages. In fact, accumulations of limonite, or bog iron, were mined as the source of ore for this country's first ironworks and for paint pigment (see Figure 8). Limonite deposits are most common in the bog regions of Connecticut, Massachusetts, Pennsylvania, New York, and elsewhere along the Appalachians. Wetlands were abundant in parts of the Tennessee Valley during past ages, and significant bog iron deposits are found in Virginia, Tennessee, Georgia, and Alabama. Although now of limited economic importance in the U.S., bog iron is still a significant source of iron ore in northern Europe.

Similarly, mixed oxides of manganese, called wad or bog manganese, are the product of less acidic wetland removal processes. Often these wad deposits also contain mixed oxides of iron, copper, and other metals.

SUMMARY

Partially because wetlands is a deliberately generic term, but more importantly because of the unifying characteristic that is embodied in a fascinating but sometimes bewildering variety of forms, wetlands span the spectrum from mangrove and cypress swamps through fresh and saltwater marshes to bogs and fens. Despite the variety, wetlands have two characteristics that are common to wetlands of most interest to us. They have soils or substrates that are saturated for long periods or much of the growing season and because of that, they have vegetative types with specialized structures, aerenchyma, cypress knees, and buttresses that transport oxygen to their roots for

Figure 8.   This small stream in the Black Hills of South Dakota was destroyed during a bog iron mining operation in the mid-1960s.

respiration, enabling these plants to grow in an otherwise hostile environment. Secondly, many elements and compounds occur in reduced states in saturated, anoxic soils causing characteristic colors and textures. Though a variety of wetlands definitions have been developed, these two attributes, saturated or hydric soils and hydrophytic vegetation are common to almost all.

THREE IMPORTANT COMPONENTS — WATER, SOIL, AND VEGETATION

HYDROLOGY

The long-term success of any wetlands restoration or creation project is, to a very large extent, dependent upon restoring, establishing, or developing and managing the appropriate hydrology. Wetlands hydrology determines abiotic factors such as water availability, nutrient availability, aerobic or anaerobic soil conditions, soil particle size and composition, and related conditions including water depth, water chemistry (pH, Eh), and water velocity. In turn, biotic components, especially plants, influence water gains through interception of precipitation, water losses through evapotranspiration, as well as depth, velocity, and circulation patterns within the system. Plants influence water movement and even depth because vegetative resistance can create a slope to the water elevation. Water may mound in upstream areas to provide the necessary head to drive water through dense stands of downstream vegetation. In rare circumstances, notably beaver and alligators, wetland animals may have significant effects on system hydrology and a large number of muskrat houses and feeding platforms in a prairie marsh or "chimneys" built by a large population of crayfish may have substantial though usually temporary effects.

Since wetlands are transitional areas between terrestrial environments and deep water aquatic systems, they are "open" systems strongly influenced by external, forcing functions such as precipitation, solar radiation, energy and nutrient inputs, and surface and groundwater flows. Wetlands are not only spatially intermediate, but they are also intermediate in terms of amounts and chemistry of water and, consequently, they are extremely sensitive to effects of the hydrologic forcing function (see Figure 1).

Hydrology modifies or determines the structure and functioning of wetlands by:

1. Controlling the composition of the plant community and thereby the animal community. Only a few of the many thousands of species of plants are able to grow in saturated or flooded soils. Of these, adaptations to inundation vary considerably, with fewer and fewer species able to survive under longer and longer periods or deeper and deeper flooding regimes. Consequently, sites with short-term and/or shallow flooding will support many different types of plants (much higher species diversity) and consequently more species of animals. The corollary of course is that areas with deep, prolonged flooding will have fewer kinds of plants and animals. However, this concept does not extend to productivity. Basic productivity may be as high or higher in the latter, even though they have much lower diversity; that is, the amount of biomass produced or

Figure 1.   An aerial view perpendicular to the direction of water flow reveals the stream-lined, lanceolate configuration of "hummock" islands shaped by hydrologic forces in the Everglades of Florida.

supported in a simple system can equal or exceed that produced in a more complex system.

2. Hydrology directly influences productivity by controlling nutrient cycling and availability, import and export of nutrients, and fixed energy supplies in the form of organic particulates and decomposition rates. Under prolonged inundation, many important nutrients are immobilized under reducing conditions in the substrate and unavailable to plants as well as separated from the water column. Periodic drying and oxidation returns these substances to active portions of the cycles within the water column and near the surface of the substrate, resulting in an explosive growth response by plants and animals. Changes in oxygen availability and concentrations caused by inundation also strongly influence decomposition rates because anaerobic rates are generally only 10% of aerobic decomposition rates. Low decomposition rates in anaerobic environments is the principal reason why many wetlands accumulate substantial quantities of partially decomposed organic material.

Inflowing surface runoff contains variable but often substantial quantities of minerals, macro- and micronutrients, and organic material that contribute greatly to high productivities of many wetlands. Conversely, surface outflows may export significant amounts of organic material, minerals, and nutrients reducing their contribution to wetlands productivity but enhancing productivity in downstream rivers or lakes. Similarly, influent groundwaters transport minerals and some nutrients into and out of the wetlands system. Hydrology in the form of circulation patterns also controls distribution of essential growth substances within the system, often enhancing spatial heterogeneity because of differential transport of nutrients into and by-products out of portions of the system. The degree of circulation also strongly influences basic productivity with stagnant water wetlands showing much lower basic productivities than flowing water or wave-influenced systems. In addition, sedimentation and erosion, depending on circulation patterns within the system, add to physical heterogeneity and consequently species richness or diversity at any one time as well as changes over time.

In summary, water moving within the system functions analogous to the bloodstream where nutrients, energy, and byproducts are physically transported, in addition to lesser movements due to concentration gradients within free waters and in the waters near the surface of the substrate. However, most wetlands are open systems much more strongly impacted by external forcing functions than the relatively closed system in the bloodstream. In wetlands with shallow impermeable substrates (clays or well-decomposed sapric peats), low hydraulic conductivities reduce the influence of this transport process to the water column and near surface regions. In fibric or poorly decomposed peat (duff and litter layers and other highly permeable substrates), the region of active cycling may extend well below the substrate surface although exchange rates obviously decrease with depth.

The end result of all factors influencing the amount of water within the system is the water budget. The overall budget merely represents the balance between all inflows and all outflows of water, but note that this definition does not include a time-dependent variable. Generally, the water budget is determined over a 1-year period, but in some cases it may be useful to estimate the water budget for shorter periods if adequate inputs are questionable due to extremes of temperature, wind, or precipitation.

Inflows

Surface and subsurface inflows as well as direct inputs result from precipitation — generally the total water equivalent of all rain, ice, and snowfall in the region; in rare instances, direct condensation on surface objects may constitute a significant portion of the total. On bare soils or lake surfaces, all of the rainfall reaches the surface. In areas with low but dense vegetation — open bogs and sedge marshes — a small percentage of the total is intercepted before reaching the surface. However, in forested swamps, the intercepted proportion may reach 30 to 40%, substantially reducing the amount of precipitation reaching the surface. Since interception is directly influenced by structure and coverage of vegetation (i.e., a triple canopy forest intercepts much more than a single canopy, low-growing marsh), as well as intensity and duration of rain events and relative humidity, interception is proportionately less during a heavy downpour than during a short-term drizzle and also lower in humid climates than in arid regions.

In addition, two abiotic factors influence actual precipitation entering a wetlands, either directly or from surface and subsurface inputs. In arid regions, it is not unusual to see virga — a column of rain below a cloud that does not reach the surface. Because of the very low relative humidity outside of the cloud, rainfall evaporates before reaching the ground and, hence, is not included in precipitation records. However, the water equivalent of snowfall is included in climatic data even though in dry, windy regions a considerable proportion of total moisture may be lost to the atmosphere through sublimation. This is an evaporative loss resulting from moisture changing directly from a solid to a gas form under subfreezing temperatures. Absence of the liquid state and generally frozen soils or ice-covered wetlands or lakes precludes any additions to soil moisture or to surface runoff that might become inputs to the wetlands. Under appropriate conditions, many centimeters of snow depth may simply disappear long before spring temperatures cause runoff from melting snow.

Few wetlands are supported by direct precipitation alone; most depend on water inputs from surface and/or subsurface flows (see Figure 2). During and immediately after a storm, nonchannelized sheet flow may bring surface water overland from surrounding higher terrain. In between storms, surface flow through established channels, streamflow, often supplies the majority of the wetlands requirements for part or all of the year. Depending on the location of the wetlands (i.e., bordering or adjacent to a stream or river, as in most bottomland hardwood swamps or enclosing and encompassing the stream, as in many marshes and bogs), the amount of inflow is influenced by the volume of streamflow. Obviously, in the latter case, water in the stream flows into the wetlands. However, in floodplain and riparian wetlands, the amount of water entering the wetlands is influenced by the duration and intensity of the rainfall, stream capacity, bank elevation, and presence or absence of oxbows or other temporary channels. Light to moderate rain may increase streamflow, but little water enters riparian wetlands until the volume exceeds the capacity of the stream and flood waters overtop banks or channel cutoffs.

Surface runoff is often the most important source of water for natural, as well as created or restored, wetlands. Surface waters also transport quantities of fixed energy and nutrients that enhance, in some cases substantially, the productivity of wetlands.

Figure 2.   Even though this thunderstorm over the Pantanal of Brazil may create heavy localized downpours, supporting water supplies originate from precipitation over wide areas running off surrounding uplands.

Surface flows can be measured in streams or predicted from watershed and climatic data. Although estimation often requires considerable information on the source watershed, methods have been developed and are presented in Chapter 8 to predict runoff amounts under different climatic, soil, and watershed conditions.

In contrast to surface flows, our understanding of and ability to estimate subsurface flows is poor; yet some natural wetlands are dependent on groundwater supplies. In many cases, wetlands occurrence is evidence of emerging groundwaters; in others, groundwaters may flow through or transit the wetlands, and still others may receive but not discharge subsurface flows and some wetlands with porous substrates add water to underground reserves. While groundwater may be critical to an individual wetlands, subsurface flows generally have small amounts of minerals with little or none of the fixed energy and nutrients brought in by surface flows. Attempts to construct wetlands that intercept groundwaters have not been very successful due to limited understanding of locations and hydraulic gradients of underground waters. Planners attempting to use groundwater sources will need to establish a sizable network of wells and monitor elevations and flow patterns through at least one abnormally wet and dry period. Conversely, sites of emerging groundwater — seeps, springs, or artesian wells — provide excellent opportunities since only a brief review of historical flows is necessary to determine source reliability and adequacy for planned wetlands sizes and configurations.

Outflows

Evapotranspiration is the combination of water that vaporizes directly from soil or water surfaces (evaporation) and the moisture that is transported through plants to vaporize into the atmosphere (transpiration). Since few surfaces (i.e., deep lakes, bare soil, or rock) lack vegetative cover, evaporation rarely adequately estimates total losses, although the standard against which other losses are compared is based on evaporative loss from a shallow water surface known as pan evaporation. More importantly, Class A pan evaporation values are used to derive the P/E ratio for any specific region. This is the ratio of total precipitation and total evaporation, generally over a period of 1 year though in some circumstances values are calculated for monthly or shorter intervals. It is important to note that the P/E ratio compensates for sublimation losses (in the evaporative component), but not for transpiration losses since both components are developed from data collected by standard rain gages and/ or standard evaporation pans. Evaporation may also be computed from prevailing radiation, temperature, wind, and relative humidity conditions, that is, the factors that influence vapor pressures at exposed surfaces and in the surrounding air.

Evapotranspiration increases with increases in exposed surface area, solar radiation, air and surface temperatures, and wind speed and decreases with relative humidity or, basically, the same factors that influence evaporation, as shown in Figure 3. However, plants have some control over transpiration and in moisture-limiting circumstances, plants can close leaf stomata, thereby reducing exposed surface area and transpiration losses. Though rarely a factor in wetlands systems, reduced soil moisture, high radiation, and other factors affecting plant physiology often activate

Figure 3.   Evapotranspiration losses from dense emergent stands are generally lower than evaporative losses from open water surfaces because of plant influences on the microclimate near the water surface within a stand of vegetation.

plant water conservation mechanisms in terrestrial environments, causing substantial reductions in transpiration losses.

Estimates of evapotranspiration from wetlands and from nonvegetated water bodies have been developed for marshes, bogs, and swamps in North Dakota, Michigan, Minnesota, New England, Florida, Utah, Nevada, and Germany. Not surprisingly, because of the difficulty of deriving accurate measurements and the variety of climates and vegetation types, the results vary. In one instance, during the growing season, a vegetated stand was believed to have lost 80% more water than nearby open water. However, most studies have shown that evapotranspiration rates from wetlands range from 30 to 90% of losses from unvegetated or open water areas. In general, it appears that annual evapotranspiration rates in wetlands average 80% of comparable losses from open water surfaces; that is, they are approximately 80% of Class A pan evaporation rates for that region. The North Dakota study (P/E ratio <1) in a dry but cool region estimated losses at 90% of pan evaporation, but studies in Nevada and Utah support the 80% figure.

For planning purposes, evapotranspiration rates from wetlands can be assumed to be 80% of Class A pan evaporation from a nearby open site; hence, wetlands evapotranspiration and lake evaporation are roughly equal since Class A pan evaporation is 1.4 times lake evaporation. On an annual basis, approximately half the net incoming solar radiation is converted to water loss. Therefore, half the net solar radiation roughly equals the annual evaporation. Not surprisingly, seasonal patterns of evapotranspiration resemble seasonal patterns of incoming radiation because wetland reflectance changes, transpiration increases and decreases, and the mulching function of litter layer increases and decreases. Combined effects result in increases during the growing season and decreases the remainder of the year. Class A pan data integrate effects of many meteorological variables and are tabulated monthly and annually in the NOAA publication *Climatological Data*.

Reduced rates may seem contradictory because of the large amount of surface area and the "pumping" effect of plant transpiration. For example, consider the total surface area exposed to the atmosphere by a dense stand of 2 to 3-m high cattail in a square meter compared to the water surface area in a square meter of open water. Intuitively, one would think that the cattail stand would have much higher losses, and that may be the case for limited periods during the growing season. However, even during the growing season, plant structure substantially reduces evaporation losses from exposed water surfaces by shading the surface, by occupying a substantial portion of surface space, and by obstructing air movement near the water's surface such that relative humidity is near saturation for some distance above the water's surface and the saturated air is not exchanged with drier air. In turn, limited air movement along the length of plant stems and leaves maintains high humidities near plant surfaces, thereby reducing transpiration losses compared to measurements obtained from a single exposed leaf. In addition, the litter layer can cause a mulching effect. These physical factors continue to influence evaporative losses long after the plant tops have ceased active growth and even after fall die-back. Similar effects have been noted in bogs and swamps. Consequently, it is not surprising that *annual* rates for evapotranspiration from a wetlands are less than regional pan evaporation rates.

Just as surface inflows are typically the most important source for wetlands, surface outflows are likely to be the major loss component. In fact, many natural wetlands in upper portions of watersheds comprise the headwaters or source for natural streams and rivers. The significance of this is apparent from an overflight of major portions of New England, the Lake States, and eastern Canada that quickly generates the impression that almost all of the waterways originate from a bog, swamp, or lake. Farther west, beaver ponds often form the headwaters of many streams throughout the Rockies and the contributions of Okefenokee Swamp to the Sewanee River, Great Swamp and the Pamlico River in North Carolina, and numerous other examples in Louisiana, Mississippi, and Florida attest to the important role of wetlands in moderating flows in many of our streams and rivers. Without the wetlands buffering effect that impedes storm flows and augments base flows, many waterways would have dramatic floodwaters after storms, followed by dry streambeds in dry intervals.

Subsurface losses are generally much less significant because most wetlands have poorly or impermeable substrates; otherwise, the wetlands would not be present. However, a few natural wetlands may intersect groundwaters such that subsurface waters flow horizontally through the wetlands or, in a few instances, surface inflows may equal or exceed subsurface losses at least during a significant portion of the year. Though not well documented, the latter type may have an important role in recharging groundwater supplies.

Combining each of the above factors in a single term develops the water budget for the wetlands:

Inputs: 1. direct precipitation
           2. surface inflows
           3. subsurface inflows
Exports: 1. surface outflows
           2. subsurface outflows
           3. evapotranspiration

Obviously, inputs must equal or exceed exports, at least on an annual basis and, importantly, during the growing season or the site will not support a wetlands system. However, if inputs exceed exports creating saturated or inundated soils that inhibit terrestrial plants for a significant portion of the growing season, the site will support a wetlands community even though the annual balance is negative.

Determining values for inputs and exports and the storage volume in the wetlands is useful because changes in water depths or elevations can then be estimated from:

$$\Delta V = V + I - E \qquad (1)$$

and

$$\Delta L = L + \Delta V/(A \times D) \qquad (2)$$

where V = volume of storage; I = inputs; E = exports; L = water level or elevation; A = area of the wetlands; D = depth.

For some wetlands projects, turnover rate or its inverse (residence time) may be important characteristics. Turnover rate (T) is simply the ratio of system volume to flow through; that is:

$$T = I/V \qquad (3)$$

where I is expressed as a quantity over a time period (i.e., cubic meters per day) and T becomes a similar time delimited value. Conversely, retention or residence time (R) becomes:

$$R = 1/T \qquad (4)$$

or

$$R = 1/(I/V) \qquad (5)$$

Suffice it to say, hydrology is the overriding factor in presence or absence of wetlands. Hydrology governs the abiotic factors which, in turn, control or influence the biotic factors that coalesce to create the form and function of complex natural systems we call wetlands. In fact, there is merit to the common belief that if you get the hydrology right, all else will follow in due time.

SOILS

Wetlands soils provide support for wetlands plants, are the medium for many chemical transformations, and are the principle reservoir for minerals and nutrients needed by plants as well as a variety of other substances. The principle difference with upland soils is an abundance of water that typically fills soil pores or void spaces, and the most important effect of water replacing air in soil voids is the isolation of the soil system from atmospheric oxygen. As a consequence, only a very thin (1 to 5 mm) boundary layer at the soil surface has adequate oxygen to maintain aerobic/oxidizing conditions and almost everything below is anaerobic/reducing. The exception is the rhizosphere, the thin film region around each root hair that is aerobic due to oxygen leakage from the rhizomes, roots, and rootlets. Shortly after a soil is flooded, the oxygen present is consumed by microbial organisms and chemical oxidation. Diffusion of oxygen through water is many orders of magnitude slower than diffusion through well-drained soils, and lower layers quickly become and remain anaerobic. The unique qualities of saturated soils result from the many interrelated physical and chemical changes that occur because of limited oxygen (anaerobic conditions) rather than from direct effects of excess water.

Wetlands soils are generally considered hydric soils in the SCS soil classification system because they are saturated for a long enough period in the growing season to develop anaerobic conditions that favor hydrophytic vegetation. Hydric soils are further divided into 1. mineral soils having less than 12 to 20% organic matter and 2. organic soils with greater than 12 to 20% organic matter. The percentage range is due to interrelated saturation and clay content factors.

In well developed wetlands, the upper layers are often organic soils or histosols, while lower layers may consist of mineral soils though the boundary is often indistinct. Tropical wetlands with high rates of decomposition may have very thin or no organic soil layers and the substrate is almost solely mineral soil. Conversely, peat layers may be many meters thick in temperate and especially cold-climate wetlands, and the underlying mineral soil is largely isolated from the wetlands systems because of the low hydraulic conductivities of well-decomposed or sapric peat.

Organic soils have a high percentage of pore spaces (>80%) and consequently higher water holding capacities than mineral soils (50%) and are described as having lower bulk densities, that is, the dry weight of a given volume of material is less. Organic soils generally have lower hydraulic conductivities than mineral soils (except clay) so that even though organic soils may contain large amounts of water, water movement through organic soils is inhibited. Although included within the organic soil type, boundaries between the surface litter/duff layer, fibric peat layers, and sapric peat layers are indistinct and flow can be rapid through the uppermost layers down to and including much of the fibric peat layer. Consequently, a peat soil may exhibit considerable lateral water flow through the upper fibric zone, but limited movement vertically into or horizontally within the lower sapric zone.

Organic soils have a greater cation exchange capacity (CEC) and the major cations are different than in mineral soils. CEC measures the soil's capacity to fix cations on exchange sites and commonly ranges from –300 to +500 in different soils. In addition, metal cations (Ca^{2+}, Mg^{2+}, Na^+) dominate in mineral soils, while H^+ dominates in organic soils.

Saturation and loss of oxygen generally causes wetlands soils to have negative redox potentials. Redox potential (Eh) measures the soil or water's capacity to oxidize or reduce chemical substances and it often ranges from –300 to +300 millivolts (mV) in wetlands soils. Oxidation is the loss of electrons and reduction is the gain of electrons. pH represents the degree of acidity or alkalinity in terms of the hydrogen ion concentration; pH of wetlands soils varies from strongly acidic (3) to strongly alkaline (11), although most wetlands soils are circumneutral.

Eh and pH conditions and the interactions of pH and Eh influence CEC, as well as many chemical and physical reactions in the soil. Typical wetlands soils may have pH of 7 S.U. and Eh of –200 mV in which case common substances typically occur in reduced forms, that is, nitrogen as N_2O, N_2, or NH_4^+, iron as Fe^{2+}, manganese as Mn^{2+}, carbon as CH_4, and sulfur as S^-. Phosphorus is not directly affected by pH or Eh but it is indirectly affected because of its association with metals that are affected and by changes in clay particle adsorption and CEC phenomena. Form changes in turn affect solubility and availability for plant uptake or reaction with other substances and the various transformations that occur in wetlands modify organic and inorganic substances, releasing some while trapping others.

After flooding, soil oxygen is quickly consumed by microbial respiration and chemical oxidation. Subsequently, anaerobic microorganisms, that are able to use substances other than oxygen as the terminal electron acceptor during respiration, soon dominate the microbial community. Importantly, anaerobic decomposition rates are only 10% of aerobic decomposition rates and frequently much lower than carbon fixation or biomass production rates (see Figure 4).

Figure 4.   Periodic drying exposes and oxidizes nutrients and other substances in bottom substrates, releasing these materials into the water column following subsequent flooding.

In summary, the loss of soil oxygen creates difficult environmental conditions for living organisms and unusual chemical conditions which in turn result in the unique attributes of wetlands soils that contribute to their functional values. The wide range of redox potentials for periodically flooded soils vs. aerobic soils is important. Wetlands are often the major reducing ecosystem on the landscape and their most important function may be as chemical transformers of nutrients and other materials. These changes may transform organic inputs to inorganic outputs, inorganic inputs to organic outputs, or any combination of the foregoing. The complex of reactions may also cause retention within the wetlands such that the system becomes a "sink" for a variety of substances.

VEGETATION

Many terms have been applied to plants growing in semi-wet to wet environments. Some labels differentiate the simpler forms, primarily algae, from higher or vascular plants — those with physical structures to transport liquid and gaseous materials. Commonly used terms include phytoplankton, vascular aquatic plant, nonvascular aquatic plant, hydrophyte, aquatic macrophyte, vascular hydrophyte, and aquatic plant. Macro simply means larger than microscopic, whereas plankton implies small and current borne (i.e., floating or suspended in the water column without rooted attachment to the substrate). Nonvascular refers to simple plants many of which are small, even individual cells, but some such as *Chara* and *Nitella* are relatively large and possess holdfasts for attachment to the substrate but lack vascularization or internal transport structures.

A number of authors have attempted to narrowly define aquatic plant, hydrophyte, etc. to differentiate terrestrial, semiterrestrial, shallow water or deep water species, but little agreement has been reached. Therefore, it seems simpler to include all of these categories in a group termed "wetlands plants" defined as plants capable of growing in an environment that is periodically but continuously inundated for more than 5 days (d) during the growing season. Obviously, this would include a few species that are primarily upland types but capable of surviving 5 d of flooding or saturated soils. It would also include those occurring in the many intermediate flooding conditions, from infrequently flooded to shallowly flooded on to the furthest extent of deep water, rooted vegetation. At the extremes, rooted vascular plants may exist in water depths of 7 to 8 meters (m) in very clear waters and a few mosses and rooted algal forms have been found as deep as 27 m. However, the vast majority of wetlands plants are limited to water depths of less than 2 m. At the drier, upper extreme, this definition would include species only flooded, and not necessarily every year, for 5 d during the growing season downward to those species present in areas with permanent water depths of 2 m.

Most attention has focused on herbaceous plants, those with soft, flexible stems compared to woody plants having rigid, persistent stems such as shrubs and trees. Herbaceous wetlands plants are divided into free floating and rooted forms and the rooted group is then subdivided into submergent, emergent, and floating-leaved types. Although infrequently included, most woody species would be considered rooted emergents.

The free-floating category includes such types as duckweeds (*Lemna*) (see Figure 5), water meal (*Wolffia*), and water ferns (*Salvinia*). Some free-floating plants (water hyacinth — *Eichhornia*) have large root systems (up to 50% of their biomass), but many salvinids and some lemnids have lost their roots and nutrients are absorbed through modified leaves. All vegetative parts of duckweeds (*Lemna*) are reduced so that they appear to be a leaf floating on the water surface, while others (*Salvinia*) have stems with sessile leaves. The productivity of floating-leaved and free-floating wetlands plants is equal to or exceeds that of emergents probably because of relatively constant and favorable environmental conditions, relatively less support and respiratory tissue and they have a considerable percentage of enclosed gas space that may enable them to trap and use CO_2 from respiration that would otherwise be lost to the atmosphere. Free-floating species have roots with numerous root hairs or modified leaves and can successfully obtain nutrients from the water column.

Emergents are the plants most characteristic of marshes — cattail (*Typha*), bulrush (*Scirpus*), rush (*Juncus*), sedges (*Carex*) but also including bog mosses such as *Sphagnum*. Typically, emergents occur in shallow waters — 5 to 30 cm. Most emergent plants have long, erect linear leaves that reduce shading while exposing a large amount of leaf area for photosynthesis and also reduce air movement near the leaves limiting moisture loss due to transpiration. In many plants, rates of photosynthesis increase up to approximately 50% of radiation maxima above which photosynthetic rates decline eventually to 0 at high levels. Many emergents have very high light saturation levels which probably contributes to their high biomass productivities. In addition, many wetlands plants are able to use the much more efficient C_4 pathway for carbon fixation during photosynthesis instead of, or in addition to, the normal C_3 pathway. Plants using the C_4 process can withdraw and use CO_2 at concentrations as low as 20 mg/L compared to lower limits of 30 mg/L for the C_3 pathway. Emergent plants obtain their nutrients from the substrate (see Figure 6).

Submerged plants depend on water pressure/buoyancy for support and their stems and leaves are thin and pliable, with aerenchymal tissue and gas-filled voids providing buoyancy. Submergents typically occur in depths of 0.5 to 1.0 m. Their leaves are either long and thin, deeply dissected along the margins, or the leaf blades are separated into leaflets. These are adaptations to maximize leaf surface area to volume ratios for survival in an environment with reduced intensities of light available for photosynthesis; as little as 2 to 5% of full sunlight in deep or murky waters. Production of submerged plants is generally low because of low light intensities under water and the low diffusion of CO_2 in water, although some submerged plants can use CO_2 from HCO_3-, respired CO_2, and CO_2 in the sediments. Submerged plants use nutrients from both the water column and substrate.

A number of submergents, the floating leaved group, have a few thin, linear underwater leaves but depend primarily on broadened or rounded floating leaves. A few species have normal underwater and floating leaves but grow above the surface with leaves typical of terrestrial plants. To further complicate identification, leaf form within a single species may vary substantially depending on whether it is growing in occasionally flooded soils, shallow waters, or in deep water. In general, floating leaved species must survive with their leaves exposed to the air on the upper side and the lower

Figure 5.   Duckweed and giant duckweed, though only a few millimeters across, form dense mats that contribute substantially to biomass productivity and host large populations of invertebrates.

Figure 6.   Emergent stands of bulrush and cattail occupy the shallow areas, with dense growths of a submergent (Sago pondweed) barely exposed in the deeper regions.

side to water, a rather extreme set of conditions. The upper surface must contend with high radiation and fluctuating temperature effects, while the lower surface has limitations on gas diffusion. Consequently, most species have rounded leaves with entire margins and a tough, leathery texture to reduce physical damage and a hydrophobic upper surface that reduces wetting. Understandably and unlike most terrestrial plants, stomata are located on the upper surface for gas exchange with the atmosphere. Long, flexible stems and leaf petioles accommodate water level fluctuations and reduce damage from wave action while exposing maximal leaf area on the water surface.

Woody species vary from low-growing shrubs to towering cypress, spruce, and cedar. Upper portions of stems and leaves are generally similar to terrestrial forms and they may be deciduous or evergreen. Differences lie in the lower portion of the stem or trunk and in root structures. Most woody wetland plants possess specialized structures — knees, adventitious roots, prop roots, lenticels, and butt swellings — to increase gas exchange between the roots and the atmosphere, as shown in Figure 7.

Anaerobic Conditions

Wetland plants often grow in substrates with inadequate concentrations of oxygen for root respiration. Most have some ability for short-term anaerobic respiration, but they grow best when oxygen is available for respiration. To overcome the oxygen limitation, wetland plants have an extensive internal lacunae system that may occupy up to 60% of the total plant volume, whereas it may be only 2 to 7% of the volume in terrestrial plants. The lacunae or aerenchyma are air spaces that allow diffusion of atmospheric gases from aerial portions of the plant into the roots. The reverse also occurs and gases formed primarily by decomposition in the substrate diffuse into the roots and subsequently into the atmosphere.

Consequently, wetland plants can satisfy the oxygen requirements of their roots by transporting oxygen from the atmosphere through the honeycomb-like lacunae down into the roots. Gas movement is believed to be primarily due to pressure gradients generated by different concentrations, but may also be influenced by temperature, relative humidity, and wind velocity differentials. In some (*Nuphar Nymphaea, Lotus, Menyanthes,* or *Typha*), air flows into young green leaves during cool night hours and is then forced downward by solar heating of those leaves during the day and returned to the atmosphere via older yellow/tan leaves.

Adaptations to flooding may also include metabolic changes. In some species, anaerobic metabolism increases to support root metabolism at the onset of flooding. Later, new root systems with highly porous structure are produced to transport oxygen for aerobic metabolism (see Figure 8). In other species, metabolism is shifted to pathways that end in nontoxic compounds (malate instead of acetaldehyde or ethanol) under flooding conditions. Since anaerobic metabolism is much less efficient than aerobic metabolism, few species depend on it for extended periods.

Oxygen transported to the roots of wetland plants can leak out of roots and oxidize the surrounding substrate. This leakage is known as radial oxygen loss (ROL). ROL creates an oxidized zone, the rhizosphere, around the rootlets, roots, and rhizomes that supports aerobic microbial populations. Aerobic microbial metabolism detoxifies

Figure 7.   Swollen bases (butt swellings) improve gas transfer between roots and the atmosphere for these cypress in a Louisiana bayou.

Figure 8.   In areas with periodic flooding, cypress "knees" increase oxygen supplies for roots growing in anaerobic environments.

potentially hazardous substances and modifies nutrients and trace organics. The juxtaposition of a thin-film aerobic region surrounded by largely anaerobic substrates is important in nitrogen, carbon, hydrogen, sulfur, and metal cycling.

Growth and Survival Limitations

Limiting factors for wetland plants are similar to those of terrestrial species, with the obvious exception of saturated soils and anoxic root environments. Basically, wetland plants require adequate nutrients, CO_2, and sunlight to carry out photosynthesis and protection from toxic substances.

In addition to water's effects on oxygen concentrations in the substrate, water depth and clarity strongly influence light availability for submergents. Flow rate impacts oxygen and nutrient availability, substrate texture and composition, and mechanical pressure on plant structures. A number of species exhibit different leaf form if the leaves are above or below the surface and leaf form variation with increasing depth or different flow rates, as well as greater root biomass with increasing velocity is typical of others. Conversely, increased flows reduce concentrations of toxic substances and import nutrients and oxygen supporting higher production rates in slowly moving waters as compared to stagnant waters.

Most wetland plants are to some extent nutrient limited, as are terrestrial plants. In general, optimal growth is limited by concentrations and availability of nitrogen and phosphate in many wetland systems. High concentrations of various salts restrict plant establishment and growth in other systems. In addition, loose textured, highly organic soils can influence rooting depth, may provide inadequate physical support for wetlands plants, and may develop low redox potentials shifting metals to soluble toxic forms (i.e., iron and manganese).

Air and water temperatures affect biochemical reactions and inhibit growth if thermal tolerances are exceeded. Many species, intolerant of low or freezing temperatures, are restricted to tropical and subtropical regions. Generally, higher temperatures promote increased production until thermal maxima are reached, after which production ceases. However, optimal temperatures for wetland plants may be as high as 28 to 32°C.

Competition with other species may be an important limiting factor for many species. For example, *Typha latifolia*, *Phragmites*, *Lythrum salicaria*, *Eichhornia*, *Myriophyllum spicatum*, *M. brasiliense*, *Pistia stratioties*, *Elodea densa*, and *Hydrilla verticillata* aggressively spread, forcing out other species and forming monotypic stands. High algal populations and dense mats of *Lemna* or other floating species can interrupt critical sunlight for submergent species or create oxygen deficiencies in the waters below. In general, floating leaved plants do not oxygenate the water as well as submergent macrophytes.

Nearly 5000 species of plants may occur in U.S. wetlands, but only a small proportion of these comprise the dominant community in the different wetland types. However, these few species not only influence hydrology and soils, they are the factory that assembles organic, living materials from nonliving substances and provides the basis for all other forms of life. Establishing and maintaining an appropriate plant community (form) is essential to generating virtually all wetlands benefits (functions).

3   NATURE'S METHODS FOR CREATING AND MAINTAINING WETLANDS

Since our objective is to create or restore wetlands, understanding natural factors that formed existing wetlands is important in evaluating the feasibility of wetlands construction in certain regions and comprehending how to design, construct, and manage a wetlands system. Doubtless we have the ability to construct some type of wetlands almost anywhere, but the continued existence of that system may be tenuous at best; it might require inordinate amounts of time and effort to maintain or it may be impossible to establish the desired type of system. Understanding the natural factors that create and maintain wetland ecosystems enhances our ability to select appropriate wetland types and to duplicate important natural processes.

GEOLOGIC: LARGE SCALE

Tectonic

Mountain-building processes originating from plate movement and collisions result in large-scale, long-term disturbances to established drainage patterns. Where uplifting outpaces stream migration, ridges frequently interrupt rivers and streams creating basins with limited outlets that flooded become initially clear mountain lakes with wetlands fringing the margins. As material eroded from adjacent highlands is deposited in the basins, the lakes gradually become shallower and eventually water depths are suitable for various forms of wetland plants in ever-increasing portions of the basin (see Figure 1).

Differential movement of large blocks along fault lines often creates long, linear basins or valleys with limited outlets, poor drainage, and abundant wetlands. As one block moves downward and adjacent blocks move relatively upward, former drainage patterns are cut off and the valley on the surface of the downward moving block may support extensive wetland environments. If fault blocks obstruct both ends or the downslope end of the valley, a very deep lake (such as Lake Baikal) may form and persist for thousands of years or, infrequently, the basin may support extensive wetlands such as the 11 million-ha wetland complex of the Pantanal in Mato Grosso do Sul, Brazil. More commonly, one or both ends are only partially blocked and drainage patterns are quickly re-established, although the substrate materials may be

Figure 1.   Upthrusting mountain ridges interrupt drainage patterns, forming lakes that gradually fill and are transformed into wetlands.

resistant to erosion resulting in meandering streams and associated wetlands, as in the Sequatchie Valley of Tennessee.

If local or regional precipitation/evaporation ratios are high and inflows relatively constant, accumulation of eroded sediments and organic materials establishes suitable conditions for bog vegetation. Accumulation of organic material accelerates in the acidic, low decomposition rate environments of bogs and eventually biological succession results in reforestation with successively drier conditions until the basin is little different from surrounding terrestrial environments.

However, if the local or regional precipitation/evaporation ratios are low, as in lower elevations of the Rocky Mountains and the Intermountain Region, water levels in the basin decline over time but are subject to extreme annual and cyclic fluctuations. In addition, accumulated salt deposits from evaporative water losses severely restrict the types of wetland plants able to survive in brackish or saline and fluctuating waters. During wet years, runoff may fill and maintain adequate water levels to sustain wetland organisms throughout the growing season. Short-term runoff in drought cycles often supports a brief flurry of biological activity, followed by long periods of dry conditions. Though not resulting from tectonic forces, the playas of the western plains (Oklahoma and Texas) have similar hydrologic cycles and biological characteristics.

Gradually over geological time, normal drainage patterns are re-established and streams and rivers erode deeper into the valley or basin floors resulting in narrow V shaped valleys with fast-flowing rivers and little wetlands habitat. Mountain lakes and wetlands are ephemeral features in the landscape, and only young or growing mountains have poorly developed drainage patterns and extensive lakes and wetlands. Compare, for example, the low number of natural lakes and wetlands in the Appalachians and Ozarks relative to the many natural lakes and wetlands in the Cascades, Sierras, and Rockies despite the fact that most of the Rocky Mountain region receives substantially less precipitation and has higher evaporation rates than the much older Appalachians or Ozarks.

Volcanos: Land Slides, Lava Flows, and Crater Lakes

Earthquakes are often associated with tectonic activities effecting smaller scale disturbances such as landslides that relocate massive amounts of earth and rock from higher regions to block or obstruct valleys and drainage patterns, as for example Earthquake Lake in Wyoming. But earthquakes and resultant landslides blocking drainage patterns and creating lakes and wetlands are also associated with volcanic activity, which may or may not have obvious ties to recent tectonic movements. Drainage obstruction from ash, pumice, and slide materials blocked existing drainages and substantially enlarged Spirit Lake during the recent eruption of Mt St. Helens in Washington. Less obviously, lava flows substantially reshape the topography and obstruct existing drainage patterns, but infrequently create wetlands since many lava flows have subterranean tunnels and older flows become extensively fragmented and the combination supports efficient drainage. Rarely are the basins (caldera) of quiescent volcanos flooded, creating deep lakes such as in Crater Lake, Oregon that have little wetland habitats.

Glaciation: Valleys and Continental

Perhaps the most significant natural force in wetlands creation in the northern hemisphere has been glaciation, both mountain glaciers and continental glaciers (see Figure 2). Even as far south as the Sierras, mountain ranges during glacial epochs of the last million years sustained similar combinations of cool temperatures and increased precipitation, resulting in accumulations of snowfall in the higher elevations. Settling and compaction over time transformed unmelted snows into substantial thicknesses of ice that slowly began moving downward under the persistent tug of gravity. Downward migration of this large unyielding mass with crushing weight, slowly transformed well-drained V-shaped valleys into U-shaped valleys and meandering streams slowly attempting to overcome blockage by mounds of debris from lateral or terminal moraines. The slow inexorable movement of this grinding, crushing mass of ice gouged out valley walls, flattened valley floors, and deposited huge quantities of earth and rock that was bulldozed ahead of the glacial front, carried within the sheet by subterranean streams or transported on the surface of the glacier. Relocation of massive amounts of soil, rock, and debris literally dammed existing streams and simultaneous deposition of ground rock materials — "rock flour," clay particles, and other fines — produced an impermeable layer sealing the bottom of flattened valleys and basins. Extensive chains of lakes and wetlands were re-created in regions that had once known wetlands, but lost them to progressive erosion and renewed drainage patterns following original mountain-building activities. Even today, glacial meltwater and rainfall runoff flowing into poorly drained U-shaped valleys supports significant wetlands throughout the Rockies as far south as central Colorado.

On high elevation plateaus and valleys without permanent snow cover and ice sheets, the combined effects of high precipitation rates, high precipitation/evaporation ratios, and poorly established drainage patterns created extensive lakes and wetlands throughout the immature Rockies and Cascades and some still exist in much older mountain regions such as the Poconos of eastern Pennsylvania and the Adirondacks in New York.

Similar climactic factors maintain extensive wetlands in a wide circumpolar band covering much of northern North America, Europe, and Asia. In North America, its southern borders are not coincidentally concurrent with the present locations of the Monogahela and Ohio Rivers in the east and the Missouri in the west. Projecting a line along the Ohio to the east coast and northward from the upper reaches of the Missouri along the western slopes of the Rockies to the Arctic Ocean bounds one of the largest expanses of wetlands in the world. A similar delimiting band along existing rivers in Europe and Asia to the Atlantic and Pacific Oceans outlines a vast wetlands region, though the southern boundary of the Asian complex has been strongly influenced by tectonic forces shaping the Himalayas. This is not to say that these regions are or were all wetlands, though a low altitude flight over major portions of Canada might generate the impression of more water than land. Though significant areas have been drained for agriculture in the Prairie provinces of Canada, most of the region to the north and east retains extensive original wetlands complexes (see Figure 3).

Figure 2.   Mountain valley and continental glaciers reshape eroded landscapes, creating broad ares with shallow relief and poor drainage that provide the flooded conditions suitable for wetland systems.

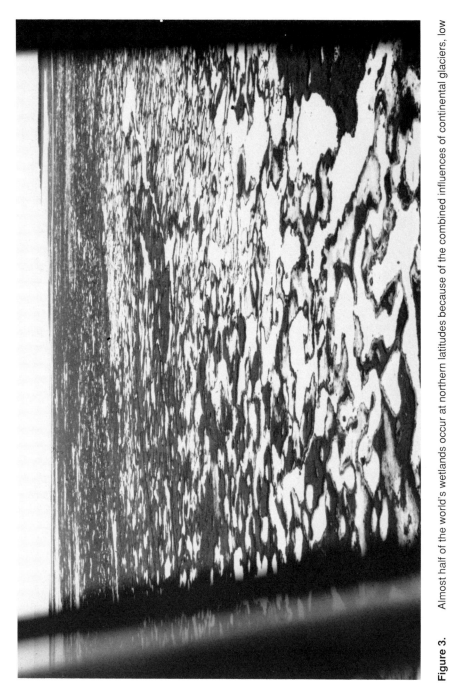

Figure 3.   Almost half of the world's wetlands occur at northern latitudes because of the combined influences of continental glaciers, low evaporation rates, and permafrost conditions.

The vast majority of the taiga and tundra regions in northern Europe, Asia, and North America consists of wetlands that, in the aggregate, represent the largest wetlands complex in the world. In fact, slightly over 45% of all natural wetlands lie between 45 and 60° North latitude. Even the combined total wetlands in the extensive deltaic and riverine swamps of the Amazon, Nile, Congo, Niger, Mekong, Ganges-Brahmaputra, Orinoco, Mississippi, and other major rivers pales in comparison to the enormous acreage of wetlands lying in a wide band circling the globe in the northern hemisphere. For example, a recent compilation found almost 400 million ha of peatlands, formed by bogs, in Canada, Alaska, Finland, Norway, United Kingdom, Iceland, and the Soviet Union with only some 52 million ha combined in the rest of the world. Of course, many tropical wetlands form little peat because of high decomposition rates, but many hectares of marshes in the northern hemisphere with limited peat deposits were also not included. Furthermore, the circumpolar northern hemisphere region is interrupted only by a few mountain ranges and of course the Atlantic, Pacific, and Arctic Oceans. Wetland expanses in the rest of the world tend to occur along widely separated river systems and lack the broad, vast continuum of wetlands common in the northern hemisphere. Though a few wetlands complexes occur in the southern extremities of South America, extreme cold in Antarctica and high temperatures with low precipitation in southern Africa and Australia preclude extensive wetlands creation or maintenance. Furthermore, these regions have not been subjected to the most significant geological force for wetlands creation.

Certainly, a much larger land mass is available in the northern hemisphere to support the shallow water conditions required for wetlands, and doubtless colder climates with high precipitation/evaporation ratios would be conducive to wetlands creation and maintenance. Yet similar climactic conditions are present in portions of Chile, Argentina, Australia, and New Zealand though largely absent in Southern Africa. Much of Africa also has the gentle relief in expansive, but well drained plains requisite for wetlands but suitable climatic regions in South America and New Zealand are occupied by rugged, high relief, mountainous terrain with limited wetlands established by tectonic forces. None of these regions has been exposed to recent influence by the major geological event in wetlands creation.

Terragenic forces from continental glaciation obliterated or relocated drainage patterns, transformed mountain ranges, and leveled broad regions throughout the circumpolar wetlands region in the northern hemisphere. Extensive climactic conditions similar to those creating and sustaining mountain glaciers formed huge sheets of ice over 2-km thick extending nearly from ocean to ocean in North America and Eurasia and reaching as far south as the present locations of the Missouri and Ohio in North America and the Tibetan plateau in Asia. Major rivers and other drainages were relocated — recent evidence suggests that the Missouri formerly flowed northward emptying into Hudson's Bay and the Arctic Ocean before its course was forced southeasterly by the Wisconsin glacier. Mountains were rounded off and valleys filled in, forming the gentle relief of present-day ranges in eastern Canada and New England that supports substantial wetlands habitats. Vast regions of interior Canada and the midwestern states were initially graded level by the bulldozing action and the crushing weight of massive sheets of ice. Pauses, advances, and retreats coupled with internal rivers within the ice sheet subsequently deposited tremendous quantities of rock, sediment, and soils forming unique patterns of low-lying hills throughout the region. Lateral moraines at the edges of these ice sheets, and terminal moraines marking southern boundaries during periods when melting paced southward movement amid numerous advances and retreats, consist of materials transported hundreds and thousands of miles before, within, and on top of the ice and deposited as extensive lines of low-lying hills. Though narrow — 3 to 30 km wide — some moraines extend linearly for hundreds of kilometers pock-marked throughout by innumerable depressions and basins of the Coteau du Missouri and other pothole regions in the Dakotas, Lakes states, and adjacent provinces in Canada. Narrower sinuous ridges (called eskars) formed by materials deposited in the beds of rivers within the ice sheet have similar rocky, hummocky terrain with countless wetlands in similar basins.

In some cases, depressions formed through differential deposition of rock and soil materials, but more commonly, present day basins represent the final resting place of ice blocks transported amid the rock, soil, and other debris ahead of, within, and beneath the glacial sheet. Although most were doubtless located in the upper regions of the 200 to 300-m of glacial till overlying much of the eastern Dakotas and western provinces, because of lower densities, melting of the few deep-lying blocks simply resulted in subsidence. Following the retreat of the ice sheet, ice blocks, near the surface of the churning layer of debris, melted leaving cavities that eroded to basins (see Figure 4). Basin bottoms were later sealed with an impermeable layer of rock flour, clay particles, and other fines eroded and wind blown from the surrounding hillsides. Hence, we find perched depressions supporting wetland communities because of a clay lens isolating surface waters from groundwaters and lacking an outlet through the surrounding hills to other surface waters and streams. Most of the myriad of pothole depressions supporting some of the most productive wetlands known are located in cavities formerly occupied by blocks of glacial ice.

In some areas, glacial meltwater flowed into large basins with limited outlets forming huge lakes — the Great Lakes and former Lake Agassiz — that covered much of the eastern Dakotas, northwestern Minnesota, and adjacent areas of Canada. Casual observations while driving west from the Red River on the Dakota-Minnesota border reveal three distinct rises, each marking the location of an ancient beachline within about 100 km. As the glacier retreated, reducing meltwater sources, the lake waters receded leaving innumerable flooded depressions from small potholes to large lakes throughout the old lake bed. Though most depressional wetlands in this area of highly fertile soils have been drained for agriculture, many still exist. Unfortunately, some of the largest — the Black Swamp of northwestern Ohio — and others in overflow regions of the Great Lakes were drained to form some of the most productive farmlands in the world. The extensive wetlands in the Hudson Bay lowlands had a similar origin but have not been significantly altered by human development.

In eastern and northern Canada, parts of New England, and much of Scandinavia, glacial scouring removed soils exposing bedrock from which huge chunks of rock were occasionally plucked out, leaving deep irregular rock-lined cavities. Though glacial meltwater filled and subsequent precipitation maintains deep lakes in many of these depressions, lake bottoms are invariably impermeable rock strata and erosion from

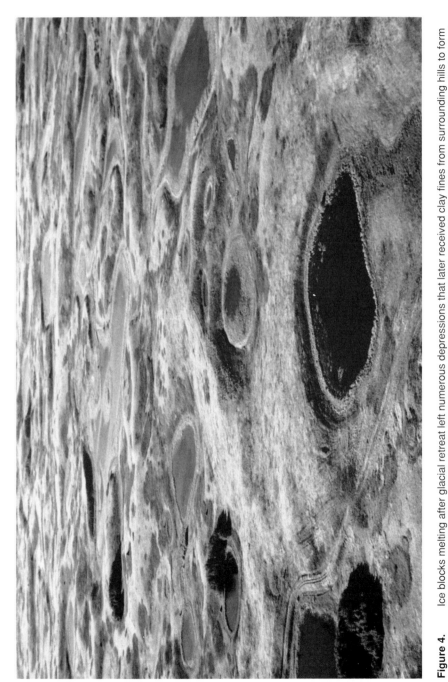

Figure 4.   Ice blocks melting after glacial retreat left numerous depressions that later received clay fines from surrounding hills to form sealing liners. The myriad potholes in the Coteau du Missouri of the Dakotas and Saskatchewan comprise the principal duck factory of North America.

largely granitic rock in the surroundings provides little nutrient input for these lakes and associated wetlands. However, bog communities have adapted to and perpetuate the acidic, low-nutrient but relatively stable waters and only the deepest lakes lack extensive fringing or neighboring bogs (see Figure 5). The progressive extension of bogs ever further onto these lakes with subsequent accumulations of deep beds of organic materials below the floating surface, gradually transforms the previous lake to the bogs and muskegs common in eastern Canada. However, these oligotrophic (and ombitrophic) lakes and wetlands are orders of magnitude less productive than the prairie potholes established on 200 to 300 m of glacial till further south and west. Lower productivities of these bedrock wetlands not only result from the very short growing season in a harsh climate, but they lack the glacial flour, clay particles, and other fines that later formed some of the world's most fertile farming soils in southern and western regions of glaciation. Acidic, low-nutrient runoff from exposed bedrock surrounding the bogs, muskegs, and taiga of eastern Canada, New England, and Scandinavia lacks the fertilizing nutrients that support the highly productive wetlands in areas of glacial deposition of the Dakotas, Lake states, Denmark, central Europe, and southern portions of the Soviet Union. Though really much more extensive — perhaps 10 to 100 times greater — the combined biomass production in these scoured out and restricted environments is considerably less than was the combined total for the original wetlands resource distributed throughout the depositional zone from Alberta to Ohio (see Figure 6).

Precipitation patterns since glacial retreat have increased the differences because the eastern bedrock regions have and continue to receive fairly high precipitation — at least adequate to support forest vegetation. High rainfall with subsequently high runoff rates, exacerbated by uptake of acidic products from tree leaf decomposition, rapidly erodes or leaches any nutrients that form on the uplands and, under low pH conditions, in lakes and wetlands, these nutrients fall unused to the bottom since plant uptake is hindered by acidic conditions. The deep clear lakes of this region are clear because the limited nutrients inflowing with runoff are quickly and irretrievably lost to deep sediments. In contrast, runoff draining prairie regions situated on hundreds of meters of glacial till has high nutrient loading in neutral or slightly alkaline waters flowing into shallow depressions where the biotic community rapidly exploits the bonanza.

The differences in runoff waters reflects soil and vegetation differences in the sources, which in turn result from not only soil or bedrock substrates but also substantial differences in the precipitation/evaporation ratios between the regions. Large portions of eastern Canada, New England, and northern Europe have high precipitation/evaporation ratios and forests dominate the vegetative communities. With low precipitation rates and/or high precipitation/evaporation ratios, grasses dominate the vegetative communities of the Plains states and the steppes of Eurasia. Not only did wetlands (and terrestrial communities) in prairie regions originate in much more fertile environments, decomposition of grasses did not yield large amounts of organic acids as does decomposition of wood and tree leaves. Consequently, the drier climate fostered establishment of highly productive wetlands by limiting invasion by forest types, thus influencing the amount and type of runoff that enters wetlands

Figure 5. Deeper depressions initially filled with glacial meltwater, are slowly transformed by floating bogs growing out from shorelines.

Figure 6.   Encroaching vegetation gradually changes the lake to a marsh, then a bog, and finally a forest.

depressions. Only very rarely is precipitation adequate to flood the prairie potholes to a high enough level to breach the surrounding landforms and initiate erosional stream formation that would eventually drain the depression.

Arctic Temperatures and Permafrost

Though precipitation rates in the high Arctic may be comparable to the southwestern deserts, only 20 to 30 cm/year, cool air temperatures reduce evaporation rates and the precipitation/evaporation ratio may be quite high. However, seemingly endless expanses of frequently more-water-than-land wetland complexes of the high Arctic are situated on infertile granitic bedrock, overall productivity is relatively low, and a large portion of the animal biomass consists of biting insects and transient birds.

The cold climates of high latitudes causes another phenomenon important to creating and maintaining tundra wetlands. Permafrost, the permanently frozen zone of soil varying from less than 1 m to many meters below the surface, prevents drainage of depressional wetlands above it during the short summers. In the absence of permafrost, many tundra wetlands would quickly lose their water to deeper ground waters and no longer support the myriads of wetlands complexes. In addition, frost heaving during cold winters continually creates small ridges and depressions with unique polygon configurations continually adding to the total wetlands environment as revealed in Figure 7.

Sea Level Changes

Sea level fluctuations over geological time have created and destroyed significant coastal and interior wetlands throughout the world. Relative emergence or subsidence of crustal plates related to plate tectonics and to glaciation, are manifested as relative changes in sea level elevations. During glacial periods, large land areas are depressed by the weight of huge ice masses while water storage in this same mass removes volume from the oceans. Consequently, sea levels may rise (relatively) along coastlines forced downward from glacial weight and concurrently fall along nonglaciated coastlines. Plate movement and subduction of one crustal plate beneath another also cause changes in relative sea level elevations on each plate, with direct impacts to coastal marshes occurring on the margins of one or both plates.

The extensive wetlands complex that occupies much of Louisiana today was created by a combination of sea level changes, erosion, and sedimentation. During a recent glacial epoch, ca. 18,000 years before present, sea level along the Gulf Coast was over 100 m lower than today's elevation. The Mississippi River cut rapidly downward, creating a wide but sharply V-shaped valley within previously deposited alluvial sediments, scoured channels across the continental shelf, and deposited sediments near the far edge of the shelf. As sea levels rose during the last 7000 years, erosion stopped and sediments were again deposited in the valley, gradually obliterating the harsh lines of the former V shape. With rising levels and valley filling, the river meandered across the new land surface, and sediments were deposited in a broad fan-shaped pattern with many constantly changing individual lobes at the valley mouth (see Figure 8).

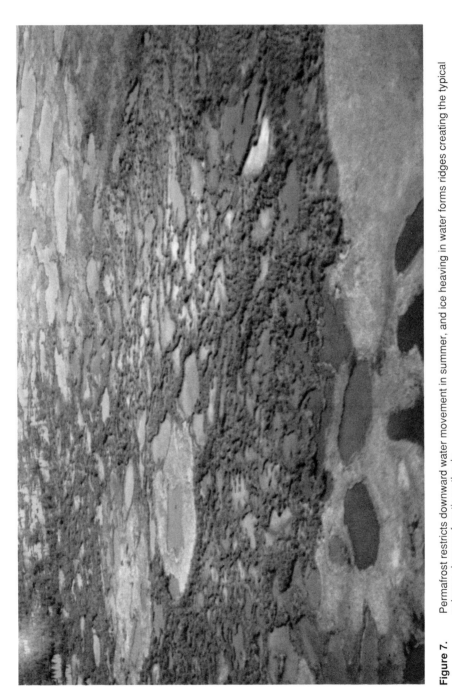

Figure 7.   Permafrost restricts downward water movement in summer, and ice heaving in water forms ridges creating the typical polygon shapes of arctic wetlands.

Figure 8.   Extensive fresh to brackish water marshes of the Mississippi Delta ebb and flow with changes in sea level and river course.

Accumulated deposits tend to restrict flow to the main channel until a breakthrough occurs, after which river flow is diverted into the new region and deposition begins in that area. As those deposits built higher or after another breakthrough at some other point, flows changed again and started the same process at another point along the coast. Once major flows no longer reached the old mouth within each lobe, fine sediments accumulated and gradually filled channels and bayous until river waters were completely cut off. During the later stages of fine sediment deposition and shortly after, the shallow water environments supporting a rich wetlands system existed. However, when deposition ceased, compaction and subsidence later caused the floor of these regions to sink, forming deep open-water bays with increasingly saline waters. In this case, changing sea level resulted in a broad valley containing a widely meandering river with countless oxbows and bayous in the upper reaches and a constantly shifting coastal depositional pattern that alternately created and destroyed wetlands (see Coleman in the references for a thorough description).

On a broader scale, rising sea levels would have mixed impacts on North American wetlands. Coastal marshes and mangrove swamps along the Gulf Coast would likely be relocated landward with overall increases in the Florida peninsula, substantial decreases in the Louisiana delta swamps and marshes, and increases along much of the East Coast (converting the Delmarva peninsula and Long Island to marsh would offset losses in the Chesapeake and Delaware Bays); but farther north, Merrymeeting Bay and Cobscook Bay in Maine would become deep-water aquatic environments. Low-lying coastal regions of Hudson's Bay and the Arctic Ocean behind the existing marshes would become roughly comparable replacements, but many delta wetlands in Alaska would be inundated and the new coast line would fringe mountain ranges only partially offset by increased wetlands acreages in interior lowlands. Further south, limited marshes in Puget Sound and other areas of the Washington and Oregon coast would be drowned, as would Humboldt Bay and other California marshes, but these would be partially offset by increases in San Francisco Bay and California's Interior Valley.

Falling sea levels would renew entrenchment of the Mississippi River, destroying most of the swamp and marsh wetlands because river flows and the water table would drop well below the land surface. Deep channels would be cut through previous deposits well out into the continental shelf and subsequent deposition would be near the limits of the shelf. Most likely, small marshes would be created in this region but the area would be much less than existing swamps and marshes. Similar patterns would be evident along much of the Gulf Coast and Florida, but concurrently falling ground water levels would eliminate most freshwater wetlands in peninsular Florida. Bay and coastal marshes behind barrier islands along the East Coast would be lost, as would substantial areas in the Chesapeake Bay and Long Island Sound. Much more extensive marshes would be created along Cape Cod and the Islands but virtually all Maine and Maritime Province marshes would be lost. Coastal wetlands of Hudson's Bay and the Arctic shores would become tundra wetlands with little overall net change, and similar changes would occur in much of Alaska. Along the West Coast, falling sea levels would eliminate virtually all existing wetlands and the steep gradient seafloor would preclude seaward replacement.

Overall, sea level alterations may increase or decrease wetland acreages depending upon the form, relief, and composition of surfaces on the landward side and gradients of sea floors on the seaward side. Major indirect effects on interior wetlands may also result from changes in groundwater elevations related to hydrostatic pressures, relocations of river deltas, and inundation or isolation of floodplains and riverine swamps. Though not as evident, the great river swamps of the Southeast and mid-Atlantic region were created by sediment deposition along river courses whose bed elevations were not much higher than sea levels. Rising sea levels would drown out these swamps and falling levels would dry them out because erosional forces would quickly deepen river beds under the effects of increased stream/river gradients.

Subsidence/Emergence: Karst Topography

Two types of extensive freshwater wetlands — the prairie potholes and the tundra wetlands — resulted from the effects of glaciation. A third major type represented by extensive areas of superficially isolated by often strongly interconnected wetlands, resulted from erosional forces on extensive layers of limestone deposits. Throughout much of Florida, the Louisville Basin of Kentucky, and the Nashville Basin of Tennessee, differential dissolution of extensive, relatively flat layers of limestone created substantial wetland acreages. As rainfall is acidified from dissolved organic materials and percolates through cracks in limestone strata, dissolution of the limestone enlarges the cracks, forming subterranean cavities and caverns. Eventually, dissolution enlarges the cavern leaving a thin roof cap that collapses downward forming a basin or pothole. Fallen roof material and washed-in debris may obstruct the drainage channel in the pothole floor and subsequent runoff floods the basin. Not surprisingly, many have circular forms and in some areas forge an irregular string of pearls. In the aggregate, the potholes or basins of karst topography originally represented substantial quantities of wetlands in the Nashville and Louisville Basins and still support significant amounts in Florida.

Though swamps with large, old hardwoods occupy many existing karst potholes in Kentucky, Tennessee, and Alabama, all of these are ephemeral systems since continued loss of limestone could re-open or create new drains at any time. In contrast, Florida potholes were eroded out during an earlier period with lower sea and groundwater levels and the waters of many today are contiguous with present groundwater tables with limited limestone erosion. High groundwater elevations also support extensive wetlands in surface depressions of these limestone layers. Though either type is relatively secure under present sea level conditions, they would be susceptible to similar threatening erosional factors if sea and groundwater levels declined significantly.

Erosion and Sedimentation

Land forms newly emerged from the sea are immediately attacked by the array of wind, water, heat, and cold erosional forces that act to reduce the new lands to sea level again. In some cases, erosion may almost offset uplifting forces and dramatically reduce elevation increases, especially in the later stages of tectonic events. In younger mountain ranges, uplifting outpaces erosion, with rapid increase in elevations. Some wetlands are created by obstructed drainages during this period, but others are caused by sediments eroded from high elevations and carried down and deposited in mountain valleys. Sediment deposition under lower gradient flows often fills the valley floor, further reducing gradients and increasing sediment deposition. Consequently, the stream or river begins to meander and undergo frequent channel alterations with cutoffs and old oxbows supporting many types of wetlands systems (see Figure 9). Cutoff oxbows and other old channels are filled with additional downward-moving sediments, but new meanders simultaneously create additional wetland habitats. Beyond the foothills and plains in the lower reaches of major rivers, sediments transported from the highest mountain peaks deposited in low stream gradient conditions similarly affect the direction and pattern of waterways. Large rivers in the lowlands frequently alter their channels, in some cases after almost every high runoff period. Extensive bottomland hardwood swamps once occupied innumerable old channels along the Mississippi, Cache, Yazoo, Tombigbee, and Appalachicola Rivers, though many have been converted to agricultural fields and levees constrain new channel changes (see Figure 10). Sediment accumulation may cause major shifts, as is presently occurring with the Mississippi and Atchafalaya; and despite concrete structures to prevent the Great River from changing its course, in time, additional sediments obstructing the present channel coupled with an extremely high flow event will likely reroute most of the flow down the Atchafalaya. The previous channel of the Mississippi and associated waterways could then become significant wetlands habitats.

Aeloian Forces: Nebraska Sandhills

Wind erosion — aeolian effects — has been a significant wetlands creation factor in a few instances. Light-weight sand and clay materials were carried for miles by prevailing winds and deposited in north central Nebraska, forming the Nebraska Sandhills region. Aeolian shaping and reforming causes hill and depression complexes typical of sand dune regions similar to glaciated prairie potholes, but with one outstanding difference. Long axis orientation of sandhill depressions represents the direction of prevailing winds rather than round forms mirroring cavities from melted ice blocks in the glaciated Dakota potholes (see Figure 11). Clay particles and other fines are carried by water to the bottom of sandhill depressions, eventually forming an impermeable layer or occasionally an isolated clay lens beneath each basin. Most larger, permanent marshes occupy linear, northwest-southeast oriented valleys intersecting groundwater above an extensive impermeable stratum underlying much of this 45,000-km^2 region. Smaller precipitation-dependent marshes have less pronounced long axis orientations, occur at higher elevations, and owe their existence to an isolated clay lens. Both types are highly productive wetlands reflecting the inherent fertility of the surrounding sand/soil materials.

Figure 9.   Cutoff oxbows and old river channels form significant wetlands in river valleys with low relief. Unfortunately, many have been ditched to improve drainage for agricultural purposes.

Figure 10.   Rivers meandering across broad flood plains create and destroy wetlands with every flood or change of course.

Figure 11.   The rich marshes of the Nebraska Sandhills benefit from continual inputs of nutrients and energy carried off surrounding hillsides.

SMALL-SCALE FACTORS

Oddly enough, a large rodent, the beaver, may have been as influential in wetlands creation in North America as any other factor. Beaver are one of the few mammals, aside from man, that have the ability to substantially modify the existing environment to suit their own requirements. Stream and river modifications by beaver dams have profoundly influenced the nature and distribution of North American wetlands. Today, beavers occur almost everywhere south of the treeline in Canada, Alaska, and the U.S., excepting only a small area in the arid southwest and most of Florida. Their present occurrence in the deep South and Texas may be due to previously depressed alligator populations and, as gator populations increase and re-occupy previous ranges, beaver distributions will likely retract northward.

At one time or another and in many cases repeatedly, beavers have influenced virtually every stream and small river throughout the continent, creating, expanding, and maintaining vital wetland habitats from coast to coast, from Mexico to the Arctic Circle and from desert valleys to mountain treelines. In aggregate, beaver established and maintained wetland habitats throughout eastern Canada and eastern U. S. (see Figure 12) and, in the Rocky Mountain chain, rival combined totals for the prairie potholes or the coastal marshes. Newly formed dams may be only 3 to 10 m long depending upon the terrain, but older colonies construct and maintain multiple dams, stair-stepping a previously free-flowing stream, maintaining extensive areas of shallow productive wetlands. In relatively level terrain, the principle dams of an older beaver complex may be 2 km long or more.

Typically, beaver erect a small dam for protection and access to food supplies shortly after colonizing a new area. As proximal food supplies are depleted, the original dam is raised and/or lengthened and additional dams are constructed upstream, downstream, or both. In steep terrain, the limit of floodable area for secure access to new food supplies is reached within a few generations and, eventually, the colony abandons the site. In flat lands, beaver colonies may occupy a drainage for hundreds of years, merely shifting up or downstream as food supplies are exhausted in one area and exploited in another. In either case, the shallow water environments created and maintained by beaver support a plethora of other wetland animals and plants, including invertebrates, fish, amphibians, reptiles, mammals, birds, and micro- and macro-phytes.

Other than a narrow band of riparian vegetation and high elevation ponds, most wetlands throughout the western mountains of North America show evidence of past or present beaver influences on natural drainage patterns (see Figure 13). Even in the older Applachians, it is not unusual to come upon a small, unusually level sedge/grass meadow adjacent to or surrounding a fast-flowing stream. In many cases, causal observation will notice a key indicator — a few old tree trunks often pines — solitary, erect, and partially buried by the deep organic sediments that underlie the meadow. Clearly, the meadow was formed by a beaver dam that flooded and killed the pines, trapped sediments, and accumulated organic materials over a number of years before the beaver left; the dam was breached and the area was reinvaded by semi-terrestrial vegetation. Core sampling often reveals this cycle has been repeated many times as

Figure 12.   The beaver lodge in the background reveals the perpetrator of an Appalachian marsh that purifies acid mine drainage waters.

Figure 13.   Beaver impoundments kill terrestrial trees, but persistent snags remain long after the beavers have left and the pond has become a wet meadow. (Photo by J. M. Hammer).

beaver colonize, deplete accessible food supplies, abandon, and then recolonize after succession has re-established suitable vegetative food supplies.

Some prairie potholes contain a large boulder with unusually smooth, almost polished sides. The rock was deposited by the glacier and may have been transported within a larger ice block that later melted, forming most of the depression. However, the polished boulder invariably lies in the deepest portion of the pothole and the most likely explanation relies on the behavior of a large herbivore. The deep portion of the depression was formed by buffalo rubbing the sides of the boulder during spring shedding. As each of many thousands of animals pushed against the rock to scrape off the accumulation of winter hair, their hooves churned the adjacent soil to fine powder and ever present winds carried the powdered soil out of the basin. Lichens and subsequent erosion have modified the rock surface, but close examination reveals a band of smoothened surface around each of these boulders corresponding to buffalo heights. Some boulders lie in small deep depressions caused by buffalo activity, while others are found in the deepest portion of a larger basin. Because of their depth and location, the latter are important refuges for wetlands species during cyclic drought periods in the pothole region.

Critical refuges for wetlands species during drought periods in the Florida Everglades, the Carolina Bays, and the Pantanal of Brazil are also created by activities of an animal. Alligators in North America and caimans in South America excavate deep and sizeable depressions — "gator holes" — during drought in many areas of the Southeast, but most noticeably in the unique bays of the Carolinas and the Everglades. These oases are doubtless formed to provide water, security, and perhaps a handy foraging area for alligators or caimans, but they also provide sanctuary for myriads of other wetland animals and plants that re-populate dessicated surroundings when the rains finally return.

MAINTENANCE

To paraphrase an old adage, the three most important factors in maintaining a wetlands are: disturbance, disturbance, and disturbance. Wetlands result from disturbance of normal drainage patterns and are quickly lost with re-establishment of normal drainage ways. Whether large scale (glaciation) or small scale (beaver dams), whether long term (tectonic forces) or short term (seasonal flooding), wetlands are absolutely dependent upon some disturbance factor for initial formation *and* for continued existence. Without disturbance — hydrologic, geologic, cyclic precipitation, fire, animal activity, etc. — wetlands are relentlessly eliminated by ecological and geological forces that gradually replace wetlands environments and denizens with successively drier, terrestrial habitats and inhabitants. Consequently, preservation of wetlands requires a flexible philosophy and active management in contrast to preservation methods suitable for forests, grasslands, or deserts. Wetlands require flooding and adequate water depths, yet continuous inundation and stable water levels may hobble the normal productivity of many wetlands (see Figure 14).

Requisite disturbances may be large scale and long term, such as changes in sea

Figure 14.   Disturbance is critical to maintaining many types of wetlands; fire often removes accumulated organic materials, restoring water depths and reversing successional changes.

levels or lake levels, or they might be large scale but short term, such as periodic droughts in the prairie pothole region. Continuous flooding over a period of years soon immobilizes essential nutrients in the substrates and concurrently washed-in sediments accumulate. Periodic drought oxidizes and recirculates nutrients and wind erosion, during severe droughts, removes accumulated sediments, resetting the basin wetlands to earlier stages of succession.

On a shorter time frame, extensive bottomland hardwood swamps occur in flood-plains because of the annual cycle of flood disturbance. Without annual flooding, these unique systems would soon become indistinguishable from the surrounding terrestrial forest. Animal activity — alligators, caimans, ungulates — disturbing wetlands often has a similar seasonality, though it is typically on a much smaller scale. Seasonal droughts enhance conditions for fires that burn accumulated organic layers, opening up and reversing succession in bogs and other peatlands. Even daily changes can be important in maintaining some wetlands. Without the twice daily flood disturbance of tides, most of our coastal marshes would cease to exist.

Wetlands are ephemeral components of the landscape formed by drainage interruptions and maintained by geological, hydrological, and biological factors that arrest or retard the impacts of other biological and geological factors that tend to transform the wetlands into a copy of its neighboring ecosystems — upland or deep water habitats. In contrast to the latter, unique, complex, and productive wetlands thrive on disturbance and change, and soon cease to exist under long-term conditions of stability.

4   WETLANDS: FUNCTIONS AND VALUES

Wetlands have both value and function and, though often used interchangeably, these terms are not synonymous. Function describes what a wetland does, irrespective of any beneficial worth assigned by man. For example, a wetlands may function by producing 100 mosquitos/m^2 or 10 muskrats/ha, purifying 5000 m^3/day of contaminated water, or by temporarily storing 100,000 m^3 of flood waters. A function is an objective process or product. A value is a subjective interpretation of the relative worth of some wetland process or product, that is, the market or recreational value of 10 muskrats or the cost for eradicating 100 mosquitos. Values can be positive or negative and they can be high or low. For example, flood storage capacity upstream from a city has high value to the residents, yet the same wetland downstream might have low value to the city because it provides them limited flood protection. Of course, the downstream wetlands could provide water purification, wildlife, recreational, or other functional values and be of high value to city residents.

Some values, the sale of muskrat pelts or veneer quality timber, have well-defined tangible parameters but others — recreational, research, and educational benefits — are intangibles that can often only be quantified by estimating replacement costs or amount and willingness to pay by users. Flood protection is intermediate, in that potential flood damages can be estimated under various flooding regimes or the cost of constructing and operating a flood storage reservoir may be taken as the value of this function of a wetland. Inconvenience or inability to use a residential yard because of mosquitos may be estimated by comparing property values and accepting the difference as the value (negative) of the nearby wetlands.

Although 10 to 15 functional values are often described for wetlands, most can be grouped into 6 major categories — life support, hydrologic modifications, water purification, erosion protection, open space and aesthetics, and geochemical storage as described below:

1. Life support: includes all types of microbial, invertebrate and vertebrate animals, and microscopic and macroscopic plants
2. Hydrologic modification: includes flood storage and conversely base flow augmentation, ground water recharge and discharge, altered precipitation and evaporation, and other physical influences on waters
3. Water quality changes: includes addition and/or removal of biological, chemical, and sedimentary substances, changes in dissolved oxygen, pH, and Eh, and other biological or chemical influences on waters

4. Erosion protection: includes bank and shoreline stabilization, dissipation of wave energy, alterations in flow patterns, and velocity
5. Open space and aesthetics: includes outdoor recreation, environmental education, research, scenic influences, and heritage preservation
6. Geochemical storage: includes carbon, sulfur, iron, manganese, and other sedimentary minerals

Of these, our information is greatest for the life support functions and values, and rapidly increasing for removal of pollutants from surface waters. Understanding and quantifying the remaining functional values remains to be accomplished and, in fact, these are most frequently estimated, rather grossly at times.

LIFE SUPPORT FUNCTIONAL VALUES

Wetlands produce many and diverse forms of life and provide habitat for many others. Life forms that must have wetlands to complete their life cycle may be termed "obligate" forms and wetlands can be said to produce these forms. Examples include fish that must have wetlands for spawning or nurseries, even though adults spend most of their lives in deeper waters; snails or clams found only in wetlands, toads that lay eggs and the larval stages live in wetlands, but adults are primarily terrestrial; and of course ducks, geese, and swans. Other "facultative" life forms use wetlands when available. Though some attribute of wetlands may substantially increase survival, growth, or population sizes, these facultative forms can survive in the absence of wetlands. For example, deer and pheasant exploit food and shelter available in wetlands, but both have thriving populations in solely terrestrial environments. However, winter cover in dry or frozen marshes strongly influences individual survival in many prairie regions. The combination of food, water, and cover needed by a species to survive and reproduce is its required habitat. Facultative forms find additional but not essential components of their habitat in wetlands, whereas obligate forms find all or critically needed components.

Wetlands are the most threatened wildlife and fishery habitats of all our natural resource base. Some 87 million ha of wetlands existed in the U.S. before colonization, but only half (approximately 44 million ha) remain. Generations of fear and antagonism toward wetlands, without understanding and appreciating the important ecological and economic benefits they provide, caused wholesale destruction during the last 300 years, most dramatically during the last 100 years. Wetlands were wastelands, obstacles to orderly development and progress, that needed to be drained, ditched, and filled to tame the wilderness. And even though misguided attitudes have changed somewhat, we are still losing 100,000 to 200,000 ha of wetlands each year in the U.S.

Historical and current wetlands losses and the importance of wetlands to fish and wildlife are reflected in the large number of threatened and endangered species that require wetlands habitats. At present in the U.S., 5 of 33 endangered mammals, 22 of 70 threatened and endangered bird species, 22 of 70 endangered and threatened reptiles, and 22 of 41 threatened and endangered fishes are dependent on wetlands or found in freshwater wetland habitats during part of their life cycle. Only a few of these have potential commercial value, but without wetlands additions to the biological diversity of our natural resources, education, recreation, and research activities would be significantly impoverished.

Wetlands are dynamic, transitional, and dependent on disturbances — the most obvious of which is fluctuating water levels. Changing water depths, either daily, seasonally, or annually, strongly influences species composition, structure, and distribution of plant communities. Varying from deep inundation to complete drying with all possible intermediates creates myriads of different zones at any one time and other combinations of zones during other times of the year. Salt and temperature gradients, tide and wave action, and pH and Eh variations structure plant communities along stratified patterns similar to well-developed forests and other complex systems.

The combined interactions of abiotic and biotic factors create a diversity and abundance of habitats that make wetlands our most important wildlife habitat. Even though wetlands only occupy a small fraction of our total land area, they support a disproportionate share of our wildlife and fish. For example, over 900 species of wildlife in the U.S. require habitat components only found in wetlands at some stage in their life cycle, with many more periodically using wetlands. Members of almost all bird groups use wetlands to some extent and one third of the species of North American birds rely on wetlands for some critical habitat component.

Microscopic Forms

Valued products generated by the life support function typically consist of larger animals and plants — crayfish and clams, sport and commercial fish, ducks and muskrats, fruits and berries, or thatching and timber. But these systems, as all other ecosystems, would cease to exist without the critical roles of innumerable forms and types of microscopic life that process and transform organic and inorganic substances supporting important biogeochemical cycles and making energy and nutrients available to all higher life forms. We are just beginning to recognize this role in another functional value — water purification. Furthermore, microbially mediated reactions in wetland systems form long-term deposits of substances containing high percentages of carbon, oxygen, and hydrogen (hydrocarbons), sulfur and metal compounds that affect global atmospheric, climatic, and hydrographic parameters in active portions of the biogeochemical cycles. In addition, many of the less obvious forms — freshwater sponges and jellyfish, orchids and fungi, algae, and insects — provide significant benefits for recreation, education, and research purposes. Finally, even the smallest virus or bacterium, the larger algae, protozoa and fungi, and on up the food chain through insects and small plants and animals provide food (nutrients and energy) for valued products, the larger animals and plants.

Invertebrates

Invertebrates also process organic materials contributing to energy and nutrient cycling but, more importantly for this function, they are the foundation of most higher

Figure 1.   Spring peepers calling from shallow wetlands are the first harbingers of spring in much of eastern North America.

food chains culminating in valued vertebrates. Wetland macroinvertebrates consist largely of annelid worms, mollusks, arthropods, and insects. Most annelids, flatworms, leeches, earthworms, nematodes, and other worms burrow into substrates or adhere to submersed aquatic vegetation, but a few are free swimming. Mollusks include snails, clams, and mussels, most of which are bottom-living or associated with vegetation and almost all are important foods for fish, salamanders, turtles, mink, otter, muskrat, raccoons and birds. Arthropods are represented by crayfish and freshwater shrimp — the most abundant and widespread of all wetland invertebrates. Both are important foods of wetland mammals, fish, and birds. Aquatic larvae of insects provide abundant foods for fish, frogs, mammals, birds, and other invertebrates. Larval forms of aquatic beetles, water striders, stoneflies, damselflies, dragonflies, springtails, mayflies, midges, mosquitoes, and other insects are also important in nutrient and energy transformations.

Vertebrates: Fish, Amphibians, Reptiles, Birds, and Mammals

Small forage fishes dominate freshwater wetlands, but the larger sport and commercial species are heavily dependent on wetland minnows for food and many also breed and spawn in wetland systems. Typical minnows include such species as fathead minnows, killifishes, top minnows, shiners, mosquito fishes, and sunfishes. Larger fish species that use wetlands diurnally or seasonally are represented by gar, bowfin, pickerel, northern pike, walleye, suckers, bullheads, carp, yellow perch, catfish,

Figure 2.   Adult toads only return to water for egg-laying, but shallow wetlands are critical to survival of their tadpole stages.

crappie, and bluegill. Larval and adult stages are important foods for birds, amphibians, reptiles, and aquatic mammals, as well as to sport and commercial fishermen (see Figure 1).

Amphibians are represented by frogs, toads, and salamanders. Many forms may be largely terrestrial, but few of the 190 species of North American amphibians do not require wetlands for at least a part of their life cycle (see Figure 2). For example, adult toads are terrestrial, but larval stages are restricted to wetlands as are many tree frogs. Bull frogs and green frogs are more closely tied to aquatic environments in all life stages; but even the terrestrial forms find critical refuge in wetlands during drought periods. Not only are they important food items for other vertebrates, they generate the night sounds that contribute to unforgettable experiences in many wetland systems and, hence, have important recreational values.

Many types of salamanders occur in wetlands but the best known, most numerous, and widespread species is the tiger salamander. During spring in some areas of the U.S., thousands cross highways enroute to shallow wetland breeding sites. Most lay eggs in wetlands, but the adult stage is largely terrestrial. However, some salamanders, mudpuppies, and water sirens, never leave wetlands because their partial metamorphosis permits them to become reproductively mature in a morphologically immature and aquatic body form.

Egg masses and tadpoles as well as adult frogs, toads, and salamanders are important foods for fish, snakes, birds, mammals, and other amphibians. The larger frogs and toads are especially nonselective and, in some areas, smaller frogs including their own species are the commonest food items.

In contrast to amphibians, reptiles lay their eggs in terrestrial environments even though many species use wetlands for food, cover, and water. However, members of the three major groups of reptiles — snakes, turtles, and alligators — often typify wetlands to many members of our society. The dark and dismal swamp is often perceived as crawling with snakes, alligators, and other dreadful creatures.

The important role of beaver in creating wetland habitat was discussed in Chapter 3. What is not as commonly known is that alligators also modify wetland habitats in ways that may be crucial to other forms of wetland wildlife. Gator "holes" dug for shelter and refuge during droughts often provide the only remaining pockets of wetland habitats for other wildlife and fish during extended droughts. Without these critical refuges, repopulation of surrounding areas at the onset of rains would be dramatically slowed due to the large distances involved. Mounds of vegetation from old alligator nests also provide resting and nesting sites for birds, loafing areas for mammals, and slightly drier points for establishment of many plants. Alligators feed on a variety of fish, birds, and mammals, in some cases so heavily that prey populations may be reduced.

Turtles are the most common, most diverse, and most visible reptiles to wetlands visitors because many species bask on logs, mounds, or other elevated structures (see Figure 3). Common turtles include painted turtles, snapping turtles, mud and musk turtles, softshell turtles, sliders, cooters, box turtles, and pond turtles. Use of wetlands varies considerably, with snapping (see Figure 4), softshell, and mud turtles being truly aquatic, emerging only to lay eggs; box and wood turtles, being largely terrestrial, enter the water only to hibernate. Although turtle eggs are important foods for many mammals, adults are largely consumers. Softshells are carnivorous, feeding largely on fish; box turtles are principally herbivorous feeding on leaves, fruit, berries, and occasionally worms and insects, and painted and snapping turtles are omnivores, dining on algae, higher plants, insects, fish, and other vertebrates.

Although a common misconception of wetlands, snakes are much more important and abundant in terrestrial systems. In fact, only garter snakes are commonly found in far northern wetlands (see Figure 5). To the south, water snakes vastly out number water moccasins, though the frequency with which they are misidentified would suggest that water moccasins are very abundant. Mud snakes and queen snakes are other common wetland snakes. Snakes feed on insects, crayfish, snails, worms, amphibians, fish, and birds and are, in turn, food for turtles and birds.

To many, the abundance and diversity of birds more correctly embodies the essence of wetlands. Freshwater wetland birds include many species of heron, egret, ibis and bitterns, ducks, geese and swans, rails, coot and gallinules, cormorants and pelicans, loons and grebes, shorebirds, cranes, ospreys, eagles, falcons and owls, and many types

Figure 3.   Owing to their sun-bathing behavior, painted turtles are often the most obvious forms of life in many wetlands.

of songbirds. In fact, representatives from almost all avian groups use wetlands, and one third of all North American birds depend directly on wetlands for some critical resource. Wetlands support many types and high numbers of birds not only because of abundant food supplies, but also because wetlands provide excellent nesting and loafing sites protected from predators. With the diversity of birds represented, virtually all other types of fish and wildlife, as well as many plants, are used by one species of bird or another. Birds, in turn, are preyed upon by snakes, turtles, other birds, mammals, and, infrequently, amphibians and fish.

Though ducks and geese often typify wetlands, other types such as grebes and loons, that build floating nests and are unable to walk on land, are more dependent on wetlands. Some birds only use wetlands occasionally; for example, peregrine falcons preying on shorebirds or ducks though their nest sites may be remote dry cliffs. Of the songbirds, a few species of wrens, blackbirds, and sparrows rarely leave wetlands, but most flycatchers, warblers, and sparrows are mainly terrestrial even though many find essential food or cover during some portion of the year in wetlands. Populations of some species may be much higher (i.e., red-headed woodpeckers and chickadees) in wooded wetlands even though terrestrial forests provide adequate habitat for their populations.

Most birds are diurnal and many congregate in large numbers forming spectacular displays. Consequently, wetlands birds have developed a broad constituency and are often thought of as the most important life support functional value for wetlands. In turn, the concern for wetland birds has done much to alter public attitudes towards wetlands.

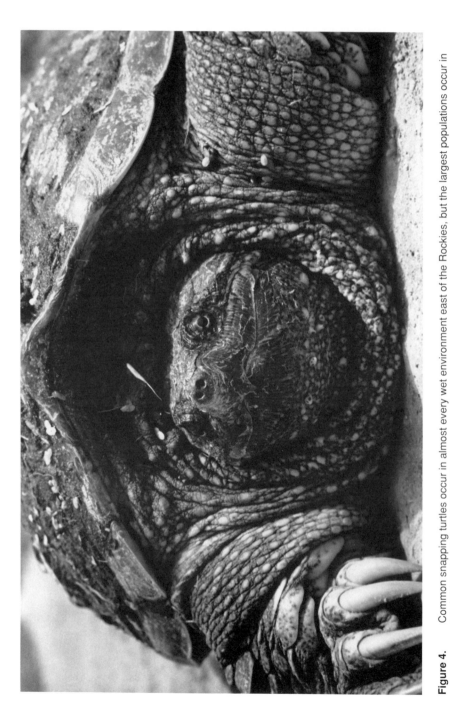

Figure 4.   Common snapping turtles occur in almost every wet environment east of the Rockies, but the largest populations occur in productive marshes and swamps.

Figure 5.   Though poisonous snakes are almost synonymous with swamps, the innocuous garter snake is the most widespread snake in North American wetlands.

Figure 6.   Raccoons were largely restricted to riparian wetlands and extensive swamps, but they have invaded the prairie marshes of the Dakotas and adjacent provinces during the last 30 years.

Few mammals are as closely tied to wetlands as are many birds, but many mammals exploit abundant food supplies or shelter during certain periods of the year (see Figure 6). Wetland-dependent mammals include muskrat, nutria, beaver, marsh rice rats, water shrews, and swamp rabbits. Muskrats inhabit almost all wetland types throughout North America, feeding primarily on emergent vegetation (notably cattail and bulrush), but also using shellfish and crayfish in many areas. Emergent vegetation supplies the raw materials for muskrat houses that are also important as nesting and loafing areas for birds, living quarters for other animals, and growing sites for semiaquatic plants.

In large rivers, muskrat and beaver dig bank dens, feed largely on deep-water shellfish and a few plants in shallow backwaters, and both feed on bark and foliage (beaver) or fruits, hard mast and invertebrates (muskrat) of terrestrial forests, especially during times of stress. But these are the exceptions. Muskrat and beaver are, for the most part, dependent on and very influential in wetlands creation and management.

Other mammals that use wetlands extensively but thrive in terrestrial habitats include lemmings, mice and voles (see Figure 7), shrews, mink, weasel, otter, raccoon, wood rats, and swamp rabbits. Wolves, coyotes, bobcats, deer, elk, and moose find food or shelter but are not restricted to wetlands. However, populations of these mammals are generally much higher in areas containing wetlands because of the additional quantities and diversity of food and shelter in wetlands (see Figures 8, 9, and 10).

Wetland plants range from microscopic forms — bacteria and minute algae — to towering cypress trees hundreds of years old. Algae include microscopic, planktonic forms (blue-green and green algae), many groups that grow attached to vegetation or other substrates (mostly green algae) as part of the periphyton community, filamentous mass forms such as *Oedogonium,* and a few macroscopic green algae such as *Chara, Cladophora* and *Nitella.* The smallest higher plants are the floating forms of duck meal (*Wolffia* and *Wolffiella*), duckweeds (*Lemna* and *Spirodela*), and the water ferns (*Azolla*). Floating forms may cover open water areas of still waters, but in areas with wind and wave action are generally restricted to protected regions along shorelines or in stands of emergent vegetation. Many floating types have high invertebrate populations associated with them and consequently provide an important food source for many wetland animals and fish. Submergent plants, typically the most important food producers for waterfowl and some fish, include pondweeds, milfoils, coontails, bladderwort, tapegrass, and widgeon grass.

Rooted floating forms include water lotus, watershield, spatterdock, and waterlily; common emergents found almost worldwide include common reed, cattail, bulrush, arrowhead, arrow arum, sweet flag, sawgrass, rushes, and spikerushes. A few of the emergents and rooted floaters produce large quantities of seed used by many animals and some (cattail, bulrush, and spatterdock), have large energy stores in tubers or roots that are heavily used by muskrat, nutria, beaver, and moose. In addition, leaves and stems of many emergent and rooted floating species are grazed by various waterfowl species, the same mammals as above, as well as deer, elk, and other large herbivores. A few emergents (wild rice, blueberries, and cranberries) produce large crops of seeds or berries used by many birds and a few mammals. Emergents also provide nesting

Figure 7. Herbivorous meadow voles and other mice form the mainstay for fox, weasel, hawks, owls, and other predators that use wetlands.

Figure 8.   Though red fox are primarily terrestrial, tracks across this frozen marsh reveal a nocturnal search for mice or rabbits in a nearby pothole.

Figure 9.   Mule and white-tailed deer find lush browsing and shelter from inclement weather in western marshes.

Figure 10.   Moose browse riparian shrubs and are often seen shoulder deep in bogs and lakes, feeding on lush wetlands vegetation.

habitat and shelter for virtually all forms of birds and mammals found in wetlands. Fish and amphibians benefit from shelter and cover provided by all plants with portions in the water column and, of course, the free-floating and rooted floating species. Many of the latter also provide fishing and hunting perches for small herons, rails, and gallinules whose elongated toes are adaptations for distributing weight, permitting these birds to walk across floating leaves (see Figure 11).

The transitional forms — smartweeds, grasses, etc. — provide shelter, nesting cover, forage, and seed crops that supply habitat for a variety of amphibians, birds, and

Figure 11.   Camouflage plumage and deliberately swaying with wind-blown plants improves a bittern's chances for avoiding detection.

mammals. Some smartweeds (*P. lapathifolium*) regularly produce abundant seed crops whereas others (*P. persicaria*) have much lower, intermittent seed production.

Shrubs and trees provide denning and nesting sites, loafing areas, hunting perches, shelter from inclement weather, and a variety of soft and hard mast to support endemic birds and mammals. Where accessible, foliage is used by deer and other herbivores but more importantly, canopy level foliage supports insect populations that in turn are food for warblers, tanagers, vireos, and other songbirds. Wading birds establish mixed nesting colonies in relatively low-growing shrubs in island situations, but prefer multilayered, near-canopy heights over land. On land, concentrated nutrient loads from bird droppings may stress trees, but colonies in flooded cypress often have high fish populations, taking advantage of the increased algae and invertebrate production beneath the colony.

Basic productivity of many wetlands far exceeds that of the most fertile farm fields (which in many cases are former wetlands). Wetlands receive, hold, and recycle nutrients continually washed from upland regions. These nutrients support an abundance of macro- and microscopic vegetation, which convert inorganic chemicals into the organic materials required — directly or indirectly — as food for animals, including man.

Although plant species diversity is often lower in wetlands, plant biomass production is greater in many wetland ecosystems than in most terrestrial habitats. Cattail marshes produce 20 to 34 metric tons/ha/year, reed marshes 15 to 27, bogs 4 to 14, wooded swamps 7 to 14, and duckweed stands 13 to 15 metric tons/ha/year. A number of pilot projects and a few operating systems employ aquatic plant production to generate methane gas or as fuel for direct heating. Peat, of course, has long been mined from wetlands for fuel (see Figure 12), and recent experiments are exploring cattail use. Reeds and bulrushes are used in various parts of the world for fuel or more commonly for thatching. Reeds are still used extensively throughout Europe and the Middle East, and bulrush and sawgrass serve a similar function in much of tropic America.

Probably the single greatest value generated by wetland plants is timber and nut production. In 1979, there were 13 million hectares of bottomland hardwood and cypress swamps in the southeastern U.S. containing 112 m³/ha of merchantable timber worth over $617/ha (or $8 billion in aggregate). Other commercially valuable products derived from plants include wild rice, blueberries, cranberries, and honey. An unusual product, medical dressings of *Sphagnum* moss, was commonly used prior to modern synthetics because of the acidic, antiseptic qualities and dried moss has been used for packing materials and gardening for many years.

Each of these wetland life forms has intrinsic values and some have tangible, quantifiable values; but the value of many is intangible and has been estimated through various means including market value, expressed willingness to pay for an activity, fixed unit day values, travel costs, or dollar amounts spent to engage in some activity. National surveys periodically conducted by the U.S. Fish and Wildlife Service estimate number of individuals engaged in outdoor activities and their expenditures. In 1980, 5.3 million hunters spent $638 million hunting migratory birds (primarily waterfowl) and 52.4 million wildlife watchers spent $9.3 billion observing and photographing waterfowl. During that same year, fur industry sales were almost $1 billion, principally from the sale of muskrat, nutria, mink, and beaver pelts. In 1979, the alligator harvest

Figure 12.   Peat (for fuel, packing material, and gardening) has long been a commercially important product of natural wetlands.

in Louisiana was worth $1.7 million and the crayfish harvest contributed $11 million to the state's economy.

Only a few attempts have been made to develop estimates of commercial value for all the life support functional products from wetland systems. A combined value assessment for a hypothetical 405-ha bottomland hardwood site in eastern Oklahoma in 1985 derived the following estimates.

	$/ha/year
Retail meat value	3.45
(Game species-rabbit, squirrel, deer, etc.)	
Hunting recreational value	25.34
Other recreational activities	31.51
Furbearer harvest	0.74
Timber harvest	87.00
Pecan harvest	116.43
Total	$264.47

In some areas, lease fees for waterfowl hunting rights or membership fees in duck clubs represent the marketable value for a tract of wetlands, but these do not reflect life support products generated from that tract. Typically, duck hunting areas in California, the lower Mississippi River Valley, along the Lake states, and in the Chesapeake Bay have lease fees ranging from $100 to $2000 per hectare and inherited memberships in prestigious hunting clubs may cost $5000 to $10,000 per year. Although waterfowl hunters place a high value on these wetlands, the desirable entity represents the functional value produced by many different wetlands, ranging from Arctic or prairie nesting grounds and mid-latitude migratory areas as well as the sought after wintering habitats.

Determining accurate, representative estimates for life support functional values has historically been difficult, largely due to the inability to employ accepted valuation methods developed for marketing. The intangible values generated by recreational activities may prove to be the most important in preserving, restoring, or creating wetlands. Although other interest groups have begun to appreciate wetland values, the historical pattern of interest and activism by hunters, wildlife watchers, photographers, and researchers created the heightened awareness in present society that led to a variety of legal and regulatory measures to protect and restore our wetland resources. Placing a market value on an evening of frog calls on a Louisiana bayou, morning clouds of geese lifting from a marsh, or afternoon dragonflies patrolling a sedge meadow may be no more likely than determining the value of a letter or visit to a congressman. However, the dedicated appreciation of those interested in the intangible life support values has made wetlands a household word and wrought surprisingly rapid changes in public attitudes towards wetlands. Some other functional values may be more easily quantified and more persuasive to decision makers but the life support functions underlie much of our historical and current interest and they deserve the credit if we do manage to reverse present trends and expand rather than reduce our nation's wetlands.

WATER PURIFICATION

Wetland ecosystems have intrinsic abilities to modify or trap a wide spectrum of water-borne substances commonly considered pollutants or contaminants. Doubtless our ancestors perceived and exploited these abilities; but in more recent times, casual observations fostered renewed interest, leading to investigations that documented changes in concentrations of various materials after processing by natural wetland systems. Much of the early work on constructed wetlands for wastewater treatment was stimulated by observing this purification phenomenon in natural wetland systems.

Many observers have noticed accelerated soil erosion after heavy rains wash across unvegetated soils and some were fortunate to encounter situations where silt-laden waters transiting natural wetlands systems were readily compared with unprocessed waters. The striking visual differences were easily verified by sampling and analysis, and the information became an important component in a communal body of knowledge on natural wetland values. Most ecologists believed this phenomenon was widespread and a few even suggested that it might occur on a large scale, though little documentation was available. I recently had the opportunity to observe an example of water quality improvements in river waters by a natural wetland system on a very large scale.

The Pantanal of western Brazil and adjacent portions of Paraguay and Bolivia is a large basin bordered by high plateaus on the east (savannah) and north (semideciduous forest) and a moderate mountain range (semideciduous forest) on the west. Runoff from these regions causes much of the 11,000,000-ha area to be flooded from December to June, and a significant but unmeasured proportion is permanently wet. Although Pantanal means marsh or swampland in Portuguese, the region is more correctly termed a large plain with a considerable amount of permanently inundated area in old river channels, meanders, lakes, smaller depressions, and potholes. In fact, an aerial overview accentuates the striking physiographic similarities with the prairie pothole region and the high Arctic in North America (absent the woody vegetation of course). An important difference is the presence of many rivers entering the Pantanal from the eastern highlands, gradually disappearing, and then reforming on the western and southern boundaries and draining off to the south. Over geological time, alluvial deposits of highlands silt has gradually transformed a flat or concave basin floor into a convex dome-like surface with higher elevations in the center and lower on the margins.

Doubtless this region provided important water improvement functions as tectonic forces created the original basin. The accumulated deposits that formed the present cross-sectional profile are dramatic evidence of previous beneficial modification of inflowing river waters; but accelerated erosion and pollution from clearing and agricultural activities and other anthropogenic sources has tremendously increased the contaminant loading of rivers draining the plateaus on the east and north. The Rio Taquiri alone carries over 30,000 metric tons of silt per day plus a variety of agrochemicals from soybean fields on the eastern plateau. Other rivers transport lower silt loads but most receive untreated sewage, industrial and mining pollution before reaching the Pantanal. For example, 1 iron ore mill used 4.8 kg of detergent per day for washing ore stacks along the Rio Correntes, gold miners use (and lose) 36,000 kg/year of mercury along the Rio Couros and the Rio Aqua Branca, and 8 alcohol distilleries (fuel) discharge 3,600,000 l/d of organic waste ("vinhoto") into rivers draining the northern plateau. The combined impact of increased pollutant loadings has caused recent hydrologic and biological changes in the upper reaches of the Pantanal that are of concern to Brazilians and conservationists worldwide.

The amazing fact is that alarmingly high concentrations of silt and pollutants in inflowing river waters are reduced to innocuous levels in waters of rivers draining the region. Examination of a topographic map (abstracted in Figure 13) provides insight into the overall processes if not the complex of mechanisms. Notice the size, especially width, of the Rio Taquiri as it drops off the plateau and enters the Pantanal on the east. A fairly wide, deep, and fast-flowing river courses out into the Pantanal and rather quickly its width, depth, and velocity are reduced. A third of the way into the Pantanal, numerous small braided streams arise, flowing perpendicularly out of the Rio Taquiri into the adjacent regions. Progressively increasing water loss with penetration into the Pantanal drastically reduces the Rio Taquiri until it almost disappears. A similar pattern is evident in the course of the Rio Aquidauana. Many of the rivers flowing into the Pantanal virtually disappear because of sheet flow and dissemination through very small waterways in this vast wetlands region (see Figure 14).

In fact, the Pantanal functions as an 11,000,000-ha sponge that absorbs inflowing waters, cleanses them of impurities, and slowly releases clean water through minor streams that aggregate into larger rivers along the southern and western boundaries. This large natural wetlands complex transforms heavily polluted influent waters into clean waters collected by the Paraguay River and used throughout much of southern South America. Not only does it provide clean water, the slow release of waters collected during the rainy season augments base flow in the Rio Paraguay during the dry half of the year, supporting adequate year-round supplies for communities, navigation, and other human and natural uses.

On a large scale as well as in local areas, natural wetlands can perform substantial improvements in water quality and quantity despite abnormal conditions. Even though the system is grossly overloaded and significant changes have occurred in wetlands of the upper Pantanal, the natural wetlands complex of the total system still provides valuable water improvements for downstream rivers. However, major land use changes in the surrounding plateaus causing accelerated pollution have only occurred within the last 5 to 10 years and alterations in the Pantanal are already evident. Undoubtedly, continued overloading will soon destroy the water improvement function, as well as important other functional values of the Pantanal in the near future. Protecting the water improvement function of the Pantanal will require extensive changes in land use, government policy, and economics of world markets. Unfortunately, other important functional values of this world-class wetlands are likely to be severely depressed before water purification is significantly damaged.

Wetlands accomplish water improvement through a variety of physical, chemical, and biological processes operating independently in some circumstances and interacting in others. Vegetation obstructing the flow and reducing the velocity enhances sedimentation and many substances of concern are associated with the sediment because of clay particle adsorption phenomena. Increased water surface area for gas exchange improves dissolved oxygen content for decomposition of organic com-

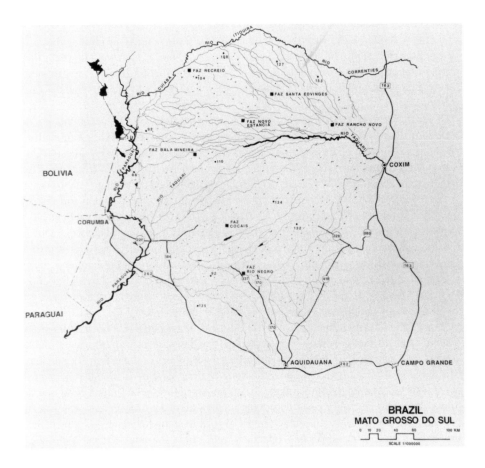

Figure 13.   A schematic diagram of the Pantanal, Mato Grosso do Sul, Brazil.

pounds and oxidation of metallic ions. But the most important process is similar to the decomposition occurring in most conventional treatment plants — only the scale of the treatment area and composition of the microbial populations is likely to be different.

In both cases, an optimal environment is created and maintained for microorganisms that conduct desirable transformations of water pollutants. Maintaining that environment in the small treatment area of a package plant requires substantial inputs of energy and labor. Wetlands systems use larger treatment areas to establish self-maintaining systems providing environments for similar microbes, but also supporting additional types of microorganisms because of the diversity of microenvironments. The latter, along with a larger treatment area, frequently provide more complete reduction and lower discharge concentrations of water-borne contaminants. Regardless, most removal or transformation of organic substances in municipal wastewaters or metallic ions in acid mine drainage is accomplished by microbes — algae, fungi, protozoa, and bacteria. Wetlands, as do conventional treatment systems, simply provide suitable environments for abundant populations of these microbial populations.

Figure 14.   The channel of the Rio Taquiri gradually disappears with distance into the Pantanal concurrent with initiation of hundreds of small streams that coalesce to form rivers on the opposite shores.

Purification Processes

Water purification functions of wetlands are dependent upon four principle components — vegetation, water column, substrates, and microbial populations. The principle function of vegetation in wetland systems is to create additional environments for microbial populations. Not only do the stems and leaves in the water column obstruct flow and facilitate sedimentation, they also provide substantial quantities of surface area for attachment of microbes (reactive surface). In addition to the microbial environments in the water column of lagoons, wetlands have much additional surface area on portions of plants within the water column. Plants also increase the amount of aerobic microbial environment in the substrate, incidental to the unique adaptation that allows wetlands plants to thrive in saturated soils. Most plants are unable to survive in water-logged soils because their roots cannot obtain oxygen in the anaerobic conditions rapidly created after inundation. However, hydrophytic or wet-growing plants have specialized structures in their leaves, stems, and roots somewhat analogous to a mass of breathing tubes that conduct atmospheric gases (including oxygen) down into the roots. Since the outer covering on the root hairs is not a perfect seal, oxygen leaks out, creating a thin-film aerobic region (the rhizosphere) around each and every root hair. The larger region outside the rhizosphere remains anaerobic, but the juxtaposition of a large, in aggregate, thin-film aerobic region surrounded by an anaerobic region is crucial to transformations of nitrogenous compounds and other substances. Wetland vegetation substantially increases the amount of aerobic environment available for microbial populations, both above and below the surface. Wetland plants generally take up only very small quantities (<5%) of the nutrients or other substances removed from the influent waters, although some systems incorporating periodic plant harvesting have slightly increased direct plant removals at considerable expense.

Recent experiments have shown that plant architecture (using temperature and relative humidity differentials between various portions of the plant) increases gas exchange beyond levels expected from passive (air tube) transport. However, attempts to compute oxygen mass introduced into the substrates by radial oxygen loss have been confounded by a number of variables that lack precise definition. Although earlier experiments suggested that plants with deep root structures had higher removal efficiencies, recent results suggest that plant species with dense, fibrous though shallow roots have lower radial oxygen loss per unit of plant biomass, but may input larger quantities of oxygen into the substrate because they tend to grow in denser stands.

Microbes (bacteria, fungi, algae, and protozoa) alter contaminant substances to obtain nutrients or energy to carry out their life cycles. In addition, many naturally occurring microbial groups are predatory and will forage on pathogenic organisms. The effectiveness of wetlands in water purification is dependent on developing and maintaining optimal environments for desirable microbial populations. Fortunately, these microbes are ubiquitous, naturally occurring in most waters and likely to have large populations in wetlands and contaminated waters with nutrient or energy sources. Only rarely, with very unusual pollutants, will inoculation of a specific type or strain of microbes be needed.

Substrates (various soils, sand, or gravel) provide physical support for plants, reactive surface area for complexing ions, anions, and some compounds, and attachment surfaces for microbial populations. Surface and subsurface water (the water column) transports substances and gases to microbial populations, carries off by-products, and provides the environment and water for biochemical processes of plants and microbes.

Invertebrate and vertebrate animals harvest nutrients and energy by feeding on microbes and macrophytic vegetation, recycling and in some cases transporting substances outside the wetlands system. Functionally, these components have limited roles in pollutant transformations, but they often provide substantial ancillary benefits (recreation/education) in successful systems. In addition, vertebrate and invertebrate animals serve as highly visible indicators of the health and well-being of a marsh ecosystem, providing the first signs of system malfunction to a trained observer.

Purification Values

Economic value for water purification has only recently been appreciated but is much less difficult to estimate than the life support functions because replacement cost is an appropriate and accepted method of estimating value. However, two approaches are possible that lead to quite different estimates. For example, a wetlands system that provides treatment of municipal wastewaters to permitted discharge standards is worth the cost of building and the annual operating cost for a conventional package plant producing similar quality effluent. The wetlands treatment system may cost $250,000 for a community of 1000 people and a comparable conventional system may cost $2.5 million. Operating costs for the wetlands system will average $10,000 per year and the conventional system will cost $100,000 per year to operate. Hence, the wetlands treating this waste is worth $2.0 million initially and $90,000 per year thereafter.

In cases where constructed wetlands have been compared to conventional package treatment plants, the cost difference favors the wetlands system and provides valid estimates of the water purification functional value of natural wetlands. However, even if the community builds a constructed wetland treatment system, the residents of the community will not value equally a comparable natural wetlands system downstream on the receiving waters. The natural system may have been performing the same function, but the community is not credited for the purification function of the natural system; hence, it is of little value to them. Our regulations require that the community treat its wastewater to specified levels for various parameters before discharging to a receiving stream or a natural wetlands. Consequently, the natural wetlands has little value to the community. In a few cases, partially treated wastewater is discharged into natural wetlands for polishing and the community manages and receives credit for water purification performed by the natural system. Doubtless the residents of these communities appreciate the water purification functional value of wetlands each time they consider higher rates for city services.

The rest of us do not receive a direct, quantifiable benefit from natural wetlands, but we do benefit in a less measurable manner. If natural wetlands remove various pollutants from surface waters that would otherwise increase the cost of treating those waters to drinking water standards, we do benefit; but determining the value of that benefit is more difficult in all but a few cases. Where acid mine drainage with high iron and manganese concentrations flows into drinking water supplies and those metals must be removed to meet drinking water standards, the cost of that removal in a conventional treatment system may be taken as the value of wetlands systems performing similar treatment.

Unfortunately, comparisons with organic contaminants are much less clear because substantial modification often occurs in intervening water bodies even though the aquatic life in those waters may be heavily impacted by the organic loading. Poorly treated wastewater may decimate aquatic life for a considerable stretch of the receiving stream or river, but gradual improvement with travel distance often produces fairly clean water that needs little, low-cost treatment to be adequate for drinking water. In this case, a wetlands between the low-quality wastewater discharge and the receiving stream would have little direct monetary value to either community though it would have considerable value to fish and other life forms and related commercial or recreational activities on the river. However, the latter value is less easily quantified and less obvious to regional residents.

Natural wetlands, especially river swamps, provide substantial benefits in treating sediment, nutrients, and agrochemicals in runoff from row crop fields and pastures protecting aquatic life in streams, rivers, and estuaries. The value of that treatment can be estimated given measurements of the annual contaminant load carried into the streams and rivers, again in terms of replacement cost (i.e., the cost of building and operating conventional facilities to accomplish the same treatment). However, the direct value is again less clear because the real worth is the value for sport and commercial fisheries or other activities based on the finfish, shellfish, bird, or mammal resources in the rivers, lakes, or estuaries. Obviously, these are valued by our society or we would not have statutory requirements on municipal or industrial discharges designed to protect these resources; but we do not have similar requirements on so-called no-point source pollution, principally agricultural wastewaters and urban stormwater runoff. Consequently, the value for treatment of dispersed but significant agricultural and urban pollution performed by millions of hectares of wetlands throughout the nation is imprecise and unappreciated (see Figure 15).

Fortunately wetlands have, for thousands of years, and continue to remove contaminants from surface waters, preserving life forms in downstream ecosystems and protecting drinking water supplies regardless of whether accurate valuations are possible or whether due appreciation develops. Wetlands can continue to perform water treatment even though so heavily impacted by other activities that most of the life support function has been lost.

HYDROLOGIC MODIFICATION

Although marshes and bogs also perform this function, floodwater modification is most often identified with bottomland hardwood swamps. Forested wetlands in river floodplains have dramatic effects on peak flows of flood waters and also on base flows during dry periods. By directly obstructing and slowing flow velocity and by acting as

Figure 15.   A large wetlands complex providing recreational and educational opportunities in Coyote Hills Regional Park, as well as urban stormwater treatment for Fremont, California.

natural reservoirs, wetlands substantially reduce the height of downstream floodwater peaks and the frequency and duration of flooding. Conversely, by retaining flood waters and slowly releasing them over extended periods, wetlands augment base flows, thereby protecting aquatic life in rivers and streams. The combination of attenuated peaks and augmented base flows results in continuous, moderate water levels in rivers influenced by wetlands as compared to elevated flood peaks, high velocities and, soon thereafter, virtually dry streambeds in rivers without the moderating influence of wetlands. Rapidly growing appreciation of this functional value has stimulated consideration of "nonstructural" flood control proposals that employ wetlands and other natural vegetation and landforms rather than earthen or concrete dams and reservoirs to protect our farmlands and cities (see Figure 16).

Only a few attempts have been made to develop quantitative estimates of the value of this important function and most use replacement cost, either the cost for flood damages or the cost for constructing and operating a flood storage dam and reservoir. A classic evaluation done by the U.S. Corps of Engineers in the St. Charles River basin in Massachusetts concluded that drainage of 3400 ha of forested wetlands would increase downstream flood damages by $17 million per year. A 40-ha swamp in Illinois was found to store over 8% of the total flood water runoff, and inflow-outflow measurements of a marsh-bog-forested wetland complex in northern Minnesota yielded estimates of flood peak reductions of 0.2 to 0.5 m at downstream communities. Other studies have shown that watersheds with 40% lake and wetlands area have flood peaks only 20% as large as watersheds with little or no wetlands area. For large floods, modeling has suggested that the flood reduction value of wetlands seems to increase with the size of the flood, the farther down the watershed the wetlands is located, and increased wetlands area. A large wetlands in the lower reaches of the watershed during a high flood event has more effect on reducing flooding than smaller wetlands in the upper reaches of the watershed during low flood events.

Groundwater recharge is a related function and directing a proportion of surface waters underground during a flood event is another means by which wetlands can reduce downstream flooding while replenishing groundwater supplies. However, the significance and magnitude of this function is poorly understood, with some wetlands undoubtedly contributing waters to underground sources but others quite obviously not. The latter is not surprising since most wetlands are underlain by impermeable materials and hydrologically separated from groundwater. Groundwater recharge has been shown to occur in isolated wetlands such as prairie marshes, cypress domes, and floodplain forests. A few studied wetlands in Wisconsin, North Dakota, and Florida had direct connections and contributed significantly to groundwaters, in one case affecting groundwater over an area of over 400 km². But other wetlands have been shown to have little influence on groundwater or, in some cases, the wetlands is present because of surfacing groundwaters. Since wetlands with substantial percolation losses may occur in the same vicinity as wetlands supported by emerging groundwaters, relationships are not well understood and rarely quantified.

Determining the value for groundwater recharge is similar to that for water purification; that is, replacement cost. If a wetlands contributes to groundwater supplies, the value of that function is the cost of supplying similar quantities of water

Figure 16.   Floodplain wetlands desynchronize and reduce peak flows and then gradually release flood waters that augment base flows, creating buffered moderate river flows during wet and dry seasons.

from some alternative source or perhaps the cost of pumping water to the surface from deep wells compared to pumping costs for shallow wells. The latter could be important in irrigation areas such as the farmlands underlain by the gradually falling Ogalla Aquifer in the western U.S.

EROSION PROTECTION

Shoreline erosion is caused by tidal currents along coasts, river currents during flooding, and wind- or boat-generated waves on lakes and reservoirs. Wetlands reduce shoreline erosion by absorbing and dissipating wave energy, by binding and stabilizing shoreline substrates, and by enhancing deposition of suspended sediments. This function is perhaps most valuable along coastlines and barrier islands, but can occasionally be important along rivers, the Great Lakes, and on reservoir shorelines. Though few studies have documented shoreline stabilization for inland waters, a number of investigations have shown that unvegetated shorelines retreat at up to four times the rate of shorelines protected by saltwater marshes. Wave-caused erosion is a serious problem on reservoir shorelines in the Upper Missouri system due to frequent high winds and long fetches across wide water bodies. The U.S.A. Corps of Engineers has conducted a number of experiments and pilot projects to vegetate these shorelines to reduce property loss and downstream dredging and channel maintenance costs.

Shoreline protection can also be valued using replacement costs; in this case, the value of property losses, storm surge flooding damages, and costs for channel maintenance. Other benefits, including reducing turbidity of waters, reducing siltation, and smothering of fishery and wildlife habitat and aesthetics, are no less important but much more difficult to quantify.

OPEN SPACE AND AESTHETICS

Many people are attracted to natural environments — those that seem untouched by man — and wetlands offer an abundance of opportunities for a "primitive" experience. Wetlands also often rank high in aesthetic value probably because of the multitude of and complex intermingling of the land-water interface that has broad appeal. Owing to the variety of waters, land forms, plants, and animals, wetlands are full of different shapes and textures, stimulating visual senses and smells and sounds to satisfy other sensual experiences. The natural appeal is understandable since wetlands are often the last areas in the landscape to undergo development.

Delayed or limited development in combination with the anaerobic, reducing environments in wetlands protects and preserves historical and anthropological resources. Ancient and historical cultures exploited the rich natural resources in wetlands, leaving behind evidence of themselves and their lifestyles that may be surprisingly well preserved in the anaerobic, acidic environments of some wetlands. Since the remains are often well preserved and the sites undisturbed by development, wetland archaeological sites are often extremely valuable study sites.

Wetlands are optimal areas for environmental education because of the facility with which important scientific principles can be demonstrated and observed. Basic principles of ecology — succession, trophic levels, food webs, and nutrient and energy cycling — are more easily shown in a small beaver pond than almost any other type of ecosystem. Of course, research on wetlands has contributed in many ways to our overall understanding of our environment (see Figure 17).

Wetlands support many types of direct recreation including hunting, trapping, fishing, wildlife watching, nature photography, berry picking, picnicking, hiking, walking, and boating, some of which have been discussed in the life support section. In a few instances, the values for these activities have been estimated with unit day, travel cost, or willingness to pay methods. Unfortunately, none of these are widely accepted in economic circles and these important functions are generally not well appreciated.

BIOGEOCHEMICAL CYCLING

Most of the functions described above are short-term, though some may extend over tens or even hundreds of years. Wetlands have another very significant but long-term functional value that is commonly overlooked and poorly understood. Wetlands function as "sinks," as traps for a variety of substances that may be immobilized in enduring wetlands or the deposits created by wetlands for long time periods from a human perspective but relatively short in a geological sense. Peat that has been mined from wetlands for ages for low quality fuel is a concentrated organic mass representing the initial stages in the formation of coal and petroleum. Most of our coal and petroleum deposits resulted from carbon fixation, or immobilization, in a wetlands system and subsequent transformations under high pressure and temperatures. On a worldwide scale, wetlands continue trapping carbon in gradually deepening deposits that will someday become coal or petroleum. Current evidence suggests that the carbon dioxide concentration in the atmosphere is increasing primarily because of the burning of fossil fuels — coal and petroleum. Since wetlands can immobilize carbon for thousands and millions of years instead of the tens or hundreds of years expected by growing more trees, would it not be less costly and more efficient to remove carbon from the atmosphere for long-term storage in restored or created wetlands than by encouraging tree planting?

Perhaps more importantly, trees and other planted vegetation do not immobilize significant quantities of sulfur, a major constituent in acidic precipitation. Sulfates that fall on or are washed into marshes, bogs, and swamps are reduced to sulfides which react with metallic ions to form insoluble substances that gradually accumulate in the organic mass that becomes peat, then coal or oil. Depending on inflow concentrations, much of the sulfur may be complexed with iron, manganese, or other metals. Again, much of the increase in atmospheric concentrations of sulfur are believed due to fossil fuel burning. As with carbon, the most effective and least costly method to remove sulfur from short-term biochemical cycling in the atmosphere-water-soil compartments is to restore and create wetlands. The processes in wetlands that formed sulfur-

Figure 17.   Wetlands in urban environments provide open space, aesthetics, and unique opportunities for education and recreational uses.

containing fossil fuel deposits are just as active today as they were millions of years ago. Should we not consider restoring and creating wetlands to enhance and improve the value of this wetlands function as a long-term solution to reduce atmospheric imbalances?

Wetlands also created many of the sedimentary mineral deposits that are important sources for our metal industries. Bog iron and wad manganese deposits were some of the earliest sources of raw materials for the iron industry in Europe and North America, though both are of limited importance today. These concentrated deposits were laid down as metallic ions and were removed from inflowing waters over thousands of years by natural wetland systems. The process is ongoing today and, in fact, increasing under deliberate efforts to build wetlands for treatment of acid mine drainage. Treatment wetlands, along with innumerable natural systems, remove dissolved metals contaminating inflowing waters, oxidize, and later reduce them to insoluble sulfide compounds. Though each new ore deposit is small, hundreds of constructed and thousands of natural systems are operating in the eastern U.S. and Canada alone and the aggregate over 50 to 100 years will be a considerable body of highly concentrated, easily processed iron ore. These shallow, easily mined deposits of rich ore will be less expensive and less environmentally damaging to use than present sources, but the complex biogeochemical processes creating iron and manganese ores also bind and remove substantial quantities of sulfur from short-term, near-surface cycles (see Figure 18).

Finally, many geochemists have noticed that mineral deposits in sedimentary rocks often have convoluted, serpentine patterns suggestive of ancient river courses. Not only ferrous materials but other metals, heavy metals, and even uranium deposits occasionally exhibit this unusual spatial distribution. Quite likely, these deposits were laid down along ancient rivers; of course, the determining factor was not the river *per se*, but the wetlands in the river's flood plain where similar biochemical processes were operating as we find in bog iron and wad manganese forming wetlands today.

Amid the worldwide concern over elevated concentrations of carbon and sulfur causing atmospheric changes that exacerbate acid precipitation and global warming, the most widely accepted solutions seem to be reducing sulfur and carbon releases and planting trees for carbon fixation. Discharge reductions may slow the trends but are unlikely to bring about significant reversals and growth rates, for even tropical trees pale in comparison with the biomass production rates of a vigorous cattail marsh. Furthermore, most tree planting campaigns employ fast-growing but relatively short-lived species that are unlikely to immobilize fixed carbon for more than 100 to 200 years.

Since measured carbon fixation rates in marshes are more than double the rates for forests, marshes immobilize sulfur and myriad other metals as well as carbon, and the accumulated materials are subtracted from short-term soil-water-atmosphere cycling, should we not give serious consideration to actively promoting marsh restoration and creation? Is it merely coincidental that acid precipitation and global warming trends accelerated concurrently with increased use of fossil fuels *and* destruction of extensive areas of natural wetlands? Perhaps we have not only increased the rate of carbon and sulfur withdrawal from long-term reservoirs (peat, coal, and oil), but simultaneously

Figure 18.   Wetlands are often the major reducing environment in the landscape; consequently, they have a major role in transforming or trapping many organic and inorganic substances.

reduced the rate of deposits into those reservoirs by eliminating the carbon and sulfur storage processes of extensive natural wetlands systems.

In the final analysis, trapping carbon, sulfur, and metallic ions in long-term storage reservoirs and eliminating their immediate impacts on the world's atmosphere and climate may be one of the most important functional values that wetlands provide to our society.

CREATION OR RESTORATION?

Restoring a natural wetland has substantial advantages over designing and building a functional wetland from scratch. If your objective is to establish a wetland in some general area, first determine whether impacted or damaged wetlands occur nearby. If you are fortunate in finding one, then evaluate the existing and try to determine previous conditions. What factor(s) has changed that degraded or eliminated the previous wetlands? Evaluate the status of the current hydrology, soils, and vegetation community in the area of interest. Is it possible or feasible to re-establish the requisite hydrology or soils? Or would it be less expensive to construct a new wetland at some other site? Examine existing land use patterns in and adjacent to the old wetland site. Would a restored wetland on this site complement or detract from adjacent land uses or would nearby activities negate your restoration efforts or detrimentally impact functional values of a restored wetland?

In the worst case, you may discover that a previous wetland has been completely filled and paved over for a parking lot. Restoration of this system may be almost as difficult as creating an entirely new wetland although past ground water levels and other hydrological factors may still be advantageous. At the other extreme, thousands of wetlands have only been ditched and drained, and installing water control structures or plugs at appropriate locations in the ditch system may be all that is needed. Simply plugging the ditches is not recommended, even though it may be initially successful, since long-term maintenance of the restored wetlands will likely require deliberate water level management to insure optimal productivities. However, in the upper Midwest, periodic drought provides the disturbance factor that enhances nutrient recycling and retards succession so that a ditch plug will likely restore the original hydrology.

Even if the site has been farmed for many years, the soils contain much of the original seed bank and an appropriate schedule for reflooding will quickly restore much of the vegetative complex. Seeds of many wetland plants exemplify maximal ability to survive under favorable conditions — in some cases for hundreds of years — and then germinate and sprout at the onset of suitable conditions. In addition, wildlife using the reflooded areas transport seeds and propagules of many wetland plants, occasionally over long distances. Inundating a previously farmed prairie pothole will restore a diverse and productive prairie marsh within 3 to 5 years, but restoring the form and function of a bottomland hardwood swamp from a soybean field may take 50 years or longer after the ditches are plugged. In both cases, much of the seed bank from the previous wetlands exists within the soils of the agricultural fields, but growth and maturation times are quite different for cattail, bulrush, and pondweeds compared to oaks, cypress, and gums.

In some cases, you may wish to expedite re-establishment of the natural vegetation or animal communities through deliberate planting or stocking. Bear in mind that the best source for planting or stocking materials is a nearby natural wetland since that stock will be adapted to the climatic and edaphic conditions of your area. Secondly, use stock from nurseries or suppliers located at the same latitude (remember that altitude translates to latitude) as your site. Do not attempt to introduce species that do not (or did not) occur in natural wetlands near your restoration site, since doing so could at best result in total planting failure and in the worst case may introduce an exotic pest species. Carefully follow the procedures in Chapter 10 for planting times, methods, and subsequent management. Remember that much of your planting or stocking effort could be accomplished much more easily through simple water level manipulation and a little patience.

Since a description of the unimpacted wetlands is unlikely to be available and projections of its current successional stage are dubious, the only practical yardstick is an undisturbed wetland in the vicinity (see Figure 1). Lacking that option, the next best alternative is to use a description of a nearby natural system in reports or journal articles. The latter source is also useful in determining exactly which successional stage of regional wetlands is most desirable and/or most easily restored.

Cost, complexity, and duration of restoration activities vary with the degree of alteration of hydrology, soils, and biotic communities from the original wetlands conditions. Generally, repair, enhancement, or restoration of an existing wetland site will be much less expensive and much more likely to succeed than creating a wetland on a terrestrial site. In either restoring or creating wetlands, the original objectives should be clearly defined and recorded, and amendments should also be recorded, since on-site modifications during construction and planting tend to be the norm rather than the exception. Many wetlands mitigation projects in recent years have been severely criticized because the original and amended plans or objectives were not adequately documented (or were unquantifiable) and the apparent failures are difficult to explain or understand. This is also true for management plans, modifications, and unusual events during the early years of operation. Failure to document a drought, deluge, storm, vandalism, etc. that may cause partial or complete failure in subsequent years results in future evaluators concluding the project managers were unable to create a wetland. In fact, deliberate amendments or circumstances beyond their control caused the failure but were not documented. Of course, goals should also be quantified to facilitate measurement, comparison, and progressive evaluations during the project.

Regardless of what type of site is selected and which restoration or creation techniques are employed, it is imperative that planners clearly and precisely determine and record the objectives of the effort. If your project will restore natural wetlands, clearly specify goals and objectives, and comparative, quantitative measures that will be used to evaluate project success. Do not forget to include a time frame in your goals,

Figure 1.   Studying a nearby natural wetlands will provide information on distribution and density to develop quantitative objectives for plant communities in the new wetlands.

but be realistic in specifying when, for example, you anticipate having 90% coverage by a selected species or a combination of species.

Objectives of the wetlands creation project should include quantitative descriptions of parameters representing the form and function of the created wetlands. Obviously, you know what you would like to see in the finished product, but will others view the new wetlands similarly? Simply stating that you will create a bog or a marsh does not provide you or others with the benchmarks to measure your progress or success. Restoring a wetland to its former, perhaps pristine state may seem desirable, but will you have comparative measurements to assess the new wetlands? It would be unusual if someone had carefully measured the structure or function of the previous system providing quantitative data for comparison with the new system. And given this unlikely happenstance, would the information be truly useful? Remember, wetlands are a transitional stage in the ecological succession of a site. Hence, careful, detailed quantitative data collected at one point in time and compared with similar information collected at another point will provide instantaneous glimpses, snapshots, of a system undergoing gradual, progressive and in many cases irreversible changes. To restore the wetland to its previous condition requires arresting or reversing the successional train of events by modifying hydrological and biological determining factors. To restore the wetland to the state that succession naturally would have established may require much less deliberate management effort if that condition can be accurately described and replicated.

Obviously, developing objectives for wetlands creation must have a slightly different basis since a wetland did not previously exist on the site. In this case, review reported descriptions of regional wetlands and visit, observe, and study nearby systems. Quantitative measurements of important elements of wetlands structure (form) at a nearby system will not only provide a yardstick for comparative evaluation of establishment techniques, but will provide considerable insight into composition of wetlands structure.

In either case, it is imperative that you develop and record quantitative parameters describing your objectives. A dense stand of cattail or some emergent plants and some open water with turtles and ducks may seem desirable but, unfortunately, none of these have the same meaning to all observers. Specifying 30% coverage by bulrush at 30 stems/m^2, 70% open water with pondweeds at 20 stems/m^2, 20 occupied nesting territories of marsh wrens, and/or 80% removal of sediment from upstream watershed runoff after 5 years are measurable parameters for readily understood and accepted comparisons. Similarly, an objective of establishing a mixed stand of willow oak (20%), cherrybark oak (20%), swamp white oak (10%), red maple (10%), tupelo gum (10%), cypress (10%), black gum (10%), willow (10%), and river birch (10%) with 1 white-tailed deer per hectare and 5 prothonatary warblers per hectare after 20 years is easily evaluated by common field methods (see Figure 2).

These examples include objectives mixing form and function, as humans all too often tend to do. Typical objectives will include a certain number of larger or more visible plants and animals and perhaps units of water recharge or purification since those obvious characteristics provide desirable benefits derived from natural wetlands systems. However, mixed objectives may be difficult to achieve and/or quantify once

Figure 2. Objectives may be defined in terms of producing a certain number of a desired species of wildlife, but care is needed since the highly productive system supporting this brood of blue-winged teal is an older, well-established system.

attained. Since wetlands functional values derive from wetlands form or structure, quantifiable objectives describing the desired form or structure are generally easier to define and measure for evaluation. Objectives should define the desired form in terms of the hydrology (flooding duration, timing, and water depths), soil or substrate composition (saturation periods, depths, organic content, and mineral composition), and plant community composition (species, growth form, size, and areal coverage). Selection of one form (structure) vs. another form will determine the type and magnitude of functions supported by the new wetlands, and form selection must be guided by careful evaluation of desired functions and requisite structure to support those functions.

Restoration objectives may include re-establishing mid or late stages of wetlands successions since much of the soil/substrate, perhaps some of the vegetation and hydrology, and likely a seed bank exist on site. However, creation objectives should describe a very early successional stage if the evaluation period is short (less than 10 years for a marsh and less than 60 years for a bog or swamp) since the complex of soils, vegetation, and hydrologic patterns necessary for many functions is unlikely to fully develop in shorter intervals. Conversely, some important functions are supported by early successional stages — young systems — and objectives defining the appropriate forms should also include descriptions of methods to periodically retard or reverse natural succession to maintain the system in an early stage.

Regardless of the functions chosen or form selected, detailed quantitative descriptions of the objectives and time interval are absolute requirements to provide guidance during development and to support subsequent evaluations.

WETLANDS FUNCTIONS AND OBJECTIVES

Development objectives typically include one or more of the commonly accepted functional values:

1. recreation
2. education
3. flood reduction
4. research
5. aesthetics
6. water purification
7. bank stabilization/protection
8. commercial products
9. base flow augmentation
10. ground water recharge

Initially, most developers will identify with one functional value and identify that as their objective; that is, create wetlands that provide wildlife habitat to support nature appreciation (bird watching, photography, etc.), sport hunting, and sport fishing. Or they may wish to have a wetlands complex intermingled with housing lots and commercial offices to enhance the diversity and attractiveness of the landscape. Others may wish to harvest fur-bearing mammals, crayfish, and lumber from their tract to supplement other income. Each of these goals will require creating a diverse, complex wetlands system. Conversely, someone interested in bank protection or water purification may be able to attain their objectives with relatively simple systems.

At this stage, the most important facet is to refine, condense, and consolidate vague, general thoughts into numerical values that can be plugged into project planning. Obviously, some objectives will require highly detailed descriptions, whereas others can be accomplished with one or two sentences. For example, a housing developer wishing to add value to a tract that will have 100 single-family housing units might include created wetlands. In this case, the objective is to provide the maximum diversity of structure and texture to increase visual stimulation, and he might specify that the wetlands will border each lot at some point, include various water depths, and have maximum irregularity in the shorelines. Other objectives would include supporting transitional, shallow water, emergent, submergent, and floating leaved herbaceous plants, as well as shrubs and trees.

In this case, quantitative objectives would include square meters of water surface, meters of shoreline length, and the exact location would be graphically depicted to insure the wetlands borders all tracts and that maximum shoreline irregularity is attained. Additional descriptions would include water depths from 0.2 m above to 1.0 m below normal water levels with locations for various depths graphically shown, along with size, location, and elevations of islands, species, and exact planting locations for each, planting densities and composition, and areal extent of each type after 3 or 5 years. Each of these is a numerical value that can be described in bid specifications and construction contracts and measured for evaluation at some future time even though quantifying and numerically evaluating the overall aesthetics of the system may be less precise.

Not surprisingly, objectives for wetlands to be used for nature appreciation or education will have similar goals since both functional values are enhanced by increasing complexity and diversity. Conversely, a system that will provide shoreline protection may only need one or two species, but the objectives should include the type, planting locations and densities, and the extent of coverage for each. At the extreme, some successful water purification wetlands consist of rectangular cells with a single emergent plant; but here too, objectives should specify size and configuration of cells, plant species and planting density, water level elevations, and rates for wastewater application and projected removal efficiencies (see Figure 3).

Most planners will begin with a single objective in mind but as they explore the available functional values, they are likely to include others. If budgetary limits do not constrain these additions, will the nature of wetlands systems preclude some objectives? Or must a sequence of priorities be established to insure that the original goal is not lost? The number of functions and derivative benefits is controlled by the wetlands structure required for each function, the type, size, and location of the wetlands required to produce those functions and benefits, and ultimately the adequacy of financial resources in accordance with site characteristics. Can a single created wetland provide more than one functional value? Of course! Can it provide all known functional values for a single population of users? Probably not, simply because one location may

Figure 3.   A constructed wetlands providing wastewater treatment and, because of its attractiveness to wildlife, important recreational benefits for Arcata, California. (Photo by R. Gearhardt.)

foster one benefit but negate another. For example, a groundwater recharge system must obviously be located in an area with highly permeable substrates, and that system is unlikely to provide appropriate water purification without jeopardizing groundwater quality. A flood water storage/buffering system must be located upstream from the community it is to protect and, without added pumping costs, using that system for water purification of the town's wastewaters is impractical. The latter objective will be less expensive if the wetlands is located downstream from the community. But either of these systems could support recreational, commercial production, shoreline protection, aesthetic, educational or research benefits, in addition to the prime goal. The ability to provide the additional values is dependent on the objectives defined and the size, diversity, and complexity of the created wetlands. As a general rule, increased numbers of functional benefits are directly related to increasing size, complexity, and diversity within the system.

A large (>5000 ha) diverse wetland could easily support the entire panoply of functional values, depending on its location with respect to other features in the landscape. Conversely, a very small but diverse system (<0.5 ha) might provide some flood retention, water purification, research, and aesthetic benefits, but its value for recreation or commercial products would be limited because of minimum area requirements for many animals and the small quantity of products (lumber, crayfish, etc.) would make management impractical.

DETERMINING FUNCTIONAL VALUES

Since wetlands functional values are governed by principles of hydrology, chemistry, and ecology, it is possible to relate expected functional values with diversity and complexity; but we need to define diversity and complexity first. In biological terms, *diversity* is defined as the number of individuals related to the number of different species represented by those individuals in a given area. For example, a system with 10,000 individuals of 10 different species would have much lower diversity than a system with 10,000 individuals of 1000 different species. Diverse systems contain fewer numbers of individuals representing larger numbers of species (i.e., 5 individuals in each of 100 different species of butterflies). Conversely, a system with low diversity might have 50 individuals in only 2 species, but it would have the same total number of butterflies and could have the same quantity of biomass.

Complexity is typically related and to some extent a corollary of diversity. A complex ecosystem has many different species at each different level, whereas a simple ecosystem has only a few species at each level. For example, penguins (tertiary consumers) feeding on small fish (secondary consumers) feeding on krill (primary consumers) feeding on plankton (producers) in the Antarctic seas represent a simple system since each level has few species even though a number of trophic levels are present. This sequence is referred to as a food chain, that is, energy in the form of food flows up a simple, straightforward pathway to the top consumer level.

Conversely, a complex system might have red and gray fox, raccoons, opossums, and bears (tertiary consumers) feeding on many different species of mice (secondary consumers) feeding on many different types of insects (primary consumers) feeding on many different types of plants (producers). In addition, the carnivores (tertiary consumers) feed directly on insects (primary consumers) and on plants (producers). Consequently, the number of components in each trophic level is much greater, as is the number of interactions and pathways between trophic levels. Energy flow may follow any of a number of different pathways from producers to the top consumer level (see Figure 4).

Two important corollaries result from these principles. In the Antarctic example, the disappearance of one component (a single species of krill) could cause collapse of the entire system since the fish and penguins dependent on krill would starve and plankton populations may increase unchecked until resources are depleted and mass die-offs occur. However, if one species of insect or one species of mouse were lost in the more complex system, the top consumers would merely shift to feeding on other species of mice, insects, or plants. The complex system has the ability to adjust to and survive disturbances (perturbations) that might destroy the simple system because the complex system has alternate pathways and considerable redundancy that create system resiliency.

Whereas the simple system might suffer major changes and may never recover, the complex system is likely to continue with only minor adjustments following a significant disturbance. Consequently, the complex system exhibits considerable stability; that is, it remains relatively unchanged or returns to the original state after experiencing a disturbance. The simple system is likely to exhibit wide fluctuations in types and sizes of populations and it may never fully recover from a disturbance. It lacks stability because it lacks the alternate pathways and redundancy incorporated in the complex system.

The geographic and climatic differences apparent in the examples of simple and complex systems is not coincidental. The simple system occurs in a harsh environment (the Antarctic), whereas the complex system is present in temperate regions. Because only a few types of plants and animals have been able to adapt to environmental extremes, unusually hot or cold, wet or dry regions often have relatively simple systems with low diversity and low complexity. Even though these simple systems have tremendous numbers of a few types of plants and animals and may be highly productive, they are much more sensitive to slight disturbances than complex systems that may have lower numbers and lower productivity.

Now we can address the original thesis — increasing complexity and diversity are directly related to the ability of a wetlands system to provide more than one functional value, as well as the ability of the system to withstand disturbing factors. If we compare the various functional values with the foregoing discussion in mind, it quickly becomes apparent that certain functions and benefits will require certain levels of diversity and complexity. In fact, we can establish a continuum describing complexity and diversity in terms of three broad categories of anticipated functional values:

water purification < hydrologic buffering < life support

with rapidly increasing diversity and complexity from left to right. If we apply the same approach to commonly used benefits we find

Figure 4.   Because most animals are opportunists, few are restricted to one trophic level. Snapping turtles fed extensively on plants and invertebrates but also prey on birds and mammals.

ground water recharge < water purification < flood reduction < shoreline protection
< bank stabilization < base flow augmentation < commercial products < research
< education < aesthetics < sport fishing < sport hunting < nature appreciation

with diversity and complexity increasing along the gradient. This scaling is based on
the minimum required to produce that function and doubtless a few of these benefits
might be reversed; for example, the differences between water purification and flood
reduction are not great, but the differences between the complexity and diversity of
wetlands generating groundwater recharge functional values compared to sport hunting functional values are substantial.

More important is the general rule that a higher system (more diverse and complex)
has the ability to also support the functions of a lower system (homogeneous and
simple), and the higher systems are inherently more stable and able to withstand
disturbances. Both concepts are important to developing and defining the objectives
for creating a wetlands system. In the first case, the more diverse and more complex
system will support the prime objectives as well as the functions of systems lower on
the scale. Secondly, the higher system is more capable of withstanding disturbance than
the lower system and is less likely to require intensive management to insure continued
operation.

In conjunction with the fact that simple systems occur in harsh environments,
complexity and diversity also serve as indicators of system health and well-being. For
example, if we establish a complex system to provide life support and flood reduction
benefits, but over time the system becomes simple and homogeneous, we should
suspect that some environmental factor is more extreme than anticipated. It may have
received more extreme or more frequent flooding than planned or some other factor
may be detrimentally impacting (constraining) the system and failure to implement
remedial measures could result in system failure with loss of flood reduction values.

Similarly, a relatively complex system created for water purification and nature
appreciation that loses diversity and complexity is likely receiving excessive wastewater loading. Harsh environmental conditions restrict the types and variety of plants and
animals that can survive with subsequent loss of the nature appreciation benefits
although the lower benefit (water purification) is still supported. Of course, environmental extremes may force the system to a simpler and simpler state, but that condition
may continue to provide water purification until some disturbance causes system
failure since simple systems lack the resiliency and stability of more complex systems.

It should be obvious from the foregoing that establishing the maximum diversity
and complexity within the constraints of budgets and site characteristics is advantageous regardless of the principle objective for creating a wetlands system. The
structure of the more diverse and more complex system will provide a greater range of
functional values and benefits, and has substantially more stability which translates
into less need for direct management to obtain the desired benefits. Initial cost savings
from creating simple systems may be vastly overshadowed by time and expense
entailed in required management over the lifetime of the project. The more diverse and
more complex systems will not only better serve the principle objective, but they will
provide ancillary benefits and reduce long-term operating costs.

Several federal, state, and even some local governmental agencies and many non-government organizations (NGOs) have information, literature, wetland areas, and staff or member expertise that is invaluable to wetlands developers. Agencies and organizations with responsibilities or interests in increasing wetlands resources should be contacted for local expertise on wetlands restoration or creation, possible cooperative ventures, or potential funding. Due to the diversity and productivity of wetlands, some aspect of wetlands ecology influences the goals or objectives of most natural resource interest groups. Local chapters, regional, or national offices may be interested in developing cooperative projects or long-term management agreements. Objectives in contacting local agencies and NGOs include:

1.   technical advice and assistance
2.   inspection and observation of existing regional wetlands
3.   cooperative development programs
4.   financial assistance
5.   long-term management

Regardless of the amount of effort expended in assembling information on wetlands from literature sources, national organizations, or agencies, its value will be overshadowed by the opportunity to visit and observe various types of local wetlands and to talk with the individuals responsible for managing them. Assembling and absorbing information from books, journals, and pamphlets is merely preparation for understanding the complexity and diversity confronting a first-time observer. Of course, a basic understanding is needed to productively discuss construction and operating methods with the responsible staff. Most wetlands managers are poorly paid and truly engaged in a labor of love. Not surprisingly, they are not reluctant to share the wealth of knowledge gained through hands-on daily experience with someone else interested in wetlands.

Reviewing the types of wetlands naturally occurring in your region will guide you in selecting the type to be constructed. If many different types of marshes, bogs, meadows, and swamps are naturally present, your options are unlimited; but if only one type — probably a marsh — is present, that is indicative of the probability of success if you attempt to construct a marsh as compared to a bog or swamp. The natural wetlands are likely located in a refuge or wildlife management area and, since it will serve as a model for the system to be constructed, spend some time observing and studying the system. Request permission to accompany the managers in their daily activities and explore in-depth how and why certain procedures have been successful and others less satisfactory in that wetlands system. Remember that the true mother lode of knowledge on wetlands management still resides on the system because most managers are too busy working with the system and handling the mountain of bureaucratic paperwork to commit their knowledge to formal reports, much less journal articles. Though rarely encountered in the field, beware the attitude that wetlands must be protected at all costs and not disturbed for any reason. That rare individual simply does not understand wetlands ecology. Conversely, you may discover that the local wetlands manager will offer assistance with your project. If you are so fortunate, remember that he/she likely is already overworked and try not to add too much to his/her load.

Nothing will replace time spent studying the natural system, reviewing progress or annual reports on it, and discussing creation and management techniques with the managers. This is an opportunity to compare goals and objectives with an actual wetlands that should be very similar to the proposed system. Do not be surprised if project goals or plans are altered after examination of an existing system. Many "failed" attempts at wetlands creation could have been avoided if the developers had assembled and used the information available from this source.

The area manager will also be aware of researchers from laboratories or universities that have investigated various aspects of model wetlands. Contact them for reports and, if possible, additional technical assistance in planning and constructing the new system.

A number of state and federal agencies and local and regional NGOs are charged with or interested in expanding the wetlands resource base. Their staff in the local or state office are knowledgable on wetlands and may have considerable information on regional systems. Some have responsibility for providing technical assistance to landowners or others interested in restoring or creating wetlands. Others may be enticed into cooperative programs to create more wetlands and a few may even have limited funds for assisting in restoration or creation projects.

Perhaps more importantly, many agencies and NGOs have permanence and their participation may be crucial to the long-term viability of the new system. Although it may be as limited as changing water levels once or twice a year or as demanding as active management of specific plant or animals populations, most wetlands require some management (see Figure 1). Without it, successional changes are likely to occur that substantially alter the system from the form that was planned and constructed. Depending on the affiliation and capabilities of the wetlands developer, soliciting participation in the planning and construction phases and developing a long-term management agreement with an agency or NGO could be critical to maintaining the wetlands and fulfilling project goals.

Developers that hope to construct a wetlands and walk away are naive and careless with their resources. In the early years, most new systems will require considerable manipulation to achieve hydrological and biological objectives. A well-established system should require much less active management, but the requirement is unlikely to disappear. Consequently, a long-term commitment by the developer or a cooperating organization is essential to project success.

Figure 1.   The National Wildlife Refuge System hosts many types of wetlands varying with regional locations that will serve as examples of the types of wetlands that can be created and maintained in that region.

In addition many of these agencies and NGOs have valuable information on potential sites that must be evaluated by wetlands developers. Details on securing specific information to assist in site selection and evaluation are presented in Chapter 8.

Initial contacts for technical or financial assistance should start with a phone call to the local office of the state or federal wildlife agency or the Soil Conservation Service. Numbers are generally listed in phone directories under "government offices." Individuals in these offices normally have daily contact with colleagues in other agencies and with the local NGOs, and can provide appropriate referrals to each.

The Conservation Directory published annually by the National Wildlife Federation (NWF) lists local, state, and regional offices of virtually all conservation organizations (government and non-government), university natural resource departments for the U.S. and Canada, as well as offices of foreign conservation agencies, pertinent periodicals, directories, and other information sources. If your local library does not have a copy, contact the NWF at 1412 Sixteenth Street, N.W., Washington, D.C., 20036-2266, (202)797-6800.

A number of federal agencies support research or management on wetlands and many have local or regional offices with experienced staff familiar with local conditions. Within the U.S. Department of the Interior, the Fish and Wildlife Service (FWS) is responsible for the National Wildlife Refuges. The vast majority of these refuges are wetlands because waterfowl hunters financially and politically supported creation of the system and establishment of many individual refuges (see Figures 2 and 3). Larger refuges have permanent staff with valuable experience in hands-on day to day restoration and management of wetlands. In addition, the FWS operates wildlife research laboratories in Laurel, MD, Jamestown, ND, and Denver and Ft. Collins, CO, with emphasis on wetlands research at the Jamestown and Laurel facilities.

The Geological Survey has conducted significant research on wetlands hydrology with staff from regional centers in Reston, VA, and Denver, CO as well as from the Geological Survey office in each state. Other Interior offices with interest or responsibility for wetlands and experienced staff include the Bureau of Land Management, the Bureau of Reclamation, and the National Park Service.

In the Department of Agriculture, the Soil Conservation Service (SCS) has staff in every state and many counties with considerable experience in watershed management and constructing and managing ponds for various uses, depending on the region of the country. Recent legislation has altered the direction of the agency and it presently has a number of staff with wetlands interest and experience. In addition, the Plant Materials Centers have conducted extensive research on a variety of species and cultivars that are useful in constructing and restoring wetlands. The Forest Service (FS) manages extensive wetlands, especially in the southeast, and its staff at the Forest Experiment Stations in Asheville, NC, and New Orleans, LA, have researched forested wetlands for many years.

All branches of the military in the Department of Defense employ professional natural resource managers on individual bases and, since some bases have extensive wetlands, many managers have considerable experience in wetlands management. In addition, the Waterways Experiment Station of the U.S. Army Corps of Engineers has investigated methods to create and restore wetlands on Corps projects throughout the

Figure 2. Many of our large "natural" wetlands — Lake Agassiz, Lake Mattamuskeet, and Okeefenokee Swamp — have been restored following drainage attempts and are currently maintained only through deliberate management efforts.

Figure 3. A large number of our refuges and wildlife management areas were established to provide habitats for ducks and geese.

U.S. for many years. With recent changes in regulations on 404 permits, the Corps has added wetlands specialists in many District offices.

For a number of years, wetlands specialists in the Environmental Protection Agency (EPA) concentrated on the agency's oversight responsibility for 404 permit applications, but recently an Office of Wetlands Protection has been created at the Washington level with staff in the regional offices, and the laboratory in Corvallis, OR, has initiated research on wetlands ecology. The Tennessee Valley Authority manages extensive wetlands and the Smithsonian Institution conducts wetlands research.

Almost all of the large federal agencies have counterparts in the governmental structure of the states with experienced staff working in the same areas on a local or statewide basis. State wildlife agencies frequently operate wildlife management areas and refuges adjacent to, but occasionally spatially separate from, the national refuges with similar experienced personnel (see Figure 4). The waterfowl management group within the state wildlife agency is an excellent contact point. Agriculture, forestry, environmental protection, and natural resource management agencies of the states have staff with extensive experience in local and regional wetlands matters.

Counterparts to federal and state agencies in the U.S. are responsible for similar programs in Canada. For example, the Canadian Wildlife Service and the provincial wildlife agencies have many years of involvement with and staff that are highly experienced in wetlands management, and cooperative research at the laboratory in Delta, Manitoba laid the groundwork for much of our information on prairie marshes.

Not surprisingly, the history of wetlands concern, management, and research in non-governmental organizations (NGOs) is at least as long and, in many cases, predates the involvement of governmental organizations. Of these, Ducks Unlimited (DU) was established by waterfowl hunters as a means to use U.S. funding to acquire and manage wetlands in Canada and, for many years, its activities were largely restricted to Canadian marshes. However, recent policy changes have created the "MARSH" program which cooperatively funds wetlands projects in the U.S. DU has a large staff throughout Canada and the U.S., many of which are former employees of state wildlife agencies or the FWS. To implement DU's acquisition and management programs, their technical staff has developed substantial expertise in hands-on wetlands restoration and creation and their fund-raising counterparts are very proficient.

The Nature Conservancy was established to acquire and protect unique natural areas and, over the years, they have developed innovative methods to finance acquisition and accept donations of land with tax or debt write-off benefits. Some areas are managed by the Conservancy but, generally, the lands are sold or otherwise transferred to a governmental organization for long-term management. Many Conservancy lands provide ideal benchmarks for comparison with other wetlands and opportunities to observe and study systems that have been little impacted by human activities. In addition, the Conservancy sponsored many of the state Natural Heritage programs that are the repositories for information on the location of unique species including those legally protected as threatened or endangered (see Figure 5).

The National Audubon Society was largely a bird-watching NGO, but it has become a very broad environmental advocacy group. It manages significant wetlands including

Figure 4.   Many large wetlands complexes are jointly owned and managed by federal and state agencies.

Figure 5.   Wetlands protection has been a high priority of the Nature Conservancy due to the large number of unique species occurring in some wetlands and the diverse, productive nature of others.

Corkscrew Swamp in southwest Florida and maintains an active research group in Tavernier, FL. Local chapters are distributed throughout the country (see Figure 6).

The National Wildlife Federation (NWF) is the largest conservation organization in the world and many of its members are dedicated to wetlands preservation or restoration. As with many NGOs, NWF has a national office and state associations composed of local clubs at the grass-roots level.

The National Wildlife Refuge Association is a small NGO but it's membership is largely retired employees of the FWS, many of which spent most of their careers managing wetlands on refuges throughout the country.

The plethora of other NGOs renders individual descriptions impractical; over 450 are listed in a recent issue of The Conservation Directory and that does not include the many state birding clubs or ornithological societies, state wildflower groups, local and state herpetological groups, and even new wetlands groups such as The Wetlands Conservancy of Tualatin, OR. Many have specialized interests unrelated to wetlands and others have narrow purposes closely tied to wetlands; still others have very broad environmental purposes, some are local or regional and others have international scope. Local and regional interest groups with broad conservation basis often have members and projects on wetlands.

Do not overlook some of the highly specialized, even single-species NGOs. For example, The Trumpeter Swan Society (TTSS), headquartered in Maple Plain, MN was organized to promote restoration and management of a single bird, but most of its members have extensive wetlands experience because swans are dependent on high-quality marshes. Consequently, the interests and policies of TTSS support restoration and management of wetlands across the nation (see Figure 7). The Raptor Research Foundation (RRF) has members managing and investigating wetlands because many raptors — osprey, bald eagles, peregrine falcons, marsh harriers, short-eared owls, snail kites, etc. — are dependent on wetlands habitats.

Professional natural resource associations such as The Wildlife Society, The Society of Wetlands Scientists, and The Ecological Society of America, etc. have broad-based memberships with expertise in all aspects of wetlands ecology. Many professional societies have regional or state chapters that are excellent sources for local expertise and some chapters are active in local preservation or restoration projects.

As knowledge of the functional values of wetlands spreads, the members of almost all NGOs concerned with natural resources realize that wetlands influence their special interest and most are becoming involved with wetlands or wetland issues to some degree.

With the rapidly growing interest and information on wetlands ecology, few project planners will be unable to assemble abundant information, considerable expertise, and perhaps cooperative funding to assist them in planning, designing, and constructing wetlands. If necessary expertise is not forthcoming from another organization, employing or contracting with a wetlands ecologist is strongly recommended to avoid the many pitfalls that have caused project failures. Contacts and agreements developed during early project phases will not only substantially increase the chances of success, but may be important in long-term viability of the new system.

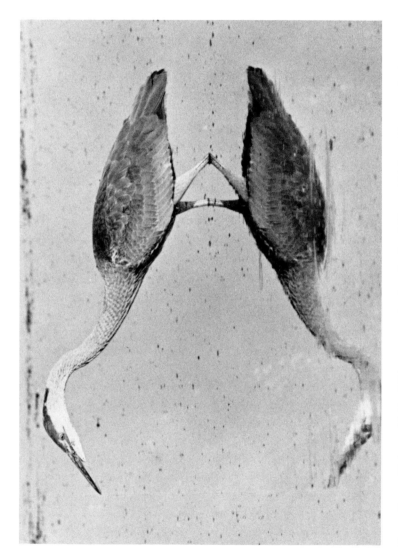

Figure 6.   Birds are highly visible components of wetlands wildlife, and bird watching organizations support wetlands protection and creation.

Figure 7.   Trumpeter swans need high-quality marshes and, consequently, The Trumpeter Swan Society is supportive of wetlands creation and protection.

SITE SELECTION AND EVALUATION

Project success is dependent on availability of a suitable site or the ability to overcome the drawbacks of a less than optimal site. The siting process provides information to identify, compare, and estimate costs of developing alternative sites at an early stage of the project. Not only will it enhance preparation of plans and drawings, but equally important is its ability to assist developers in avoiding impractical or very costly sites or at least select one with full knowledge of the consequences. Finally, site evaluation initiates the process of developing a project budget since much of the information assembled during siting relates to costs for land, construction needs, or operating parameters.

Wetlands design, construction, and operation are facilitated by identifying potential problems and opportunities early in the project with careful site evaluation. Site selection and evaluation is a systematic, reiterative process of collecting and analyzing information, modifying plans, identifying, collecting and analyzing more data, modifying plans, and repeating the cycle until the initial concept is polished into a preliminary plan that becomes the basis for the design. Siting summarizes information on topographical, geological, hydrological, soils, land ownership and use, climatic, biological, and regulatory factors that may influence construction, operation, and impacts of the proposed system. It provides baseline data for evaluating alternative sites, choosing compatible designs, assessing important components, drafting concept drawings, and outlining construction and operating plans.

Much of the discussion in this chapter is directed toward siting a created wetlands, but most factors pertinent to created wetlands sites are also relevant to evaluating the potential for restoring a wetland site. The most important single difference is the site of a restored wetland already exists and comparative evaluation of sites is meaningless or, at most, a comparison of different sites with damaged, impacted, or otherwise modified wetlands systems to select the site with the highest probability for success.

This does not mean that projects to restore wetlands will not need to conduct site investigations and evaluations (see Figure 1). On the contrary, project planners will need to acquire much the same information as those working with created wetlands, but the objectives may be slightly different. To restore wetlands, one must first determine what type of wetlands previously existed on the site. What was the form or structure and what functions were performed by the previous wetlands? What were the hydrologic and soil conditions that supported a wetlands complex on this site? What

Figure 1.   Simply plugging the ditch may not restore this bottomland hardwood swamp because increased erosion from surrounding croplands has modified channels and excessive flooding has damaged the forest.

was the composition of the plant and animal communities? Few will be fortunate to have accurate quantitative, historical data on either the form or the function of the previous wetlands. Most likely, casual descriptions and investigations of nearby healthy systems must be relied upon, in which case, detailed, extensive investigations of a neighboring system could be more lengthy and costly than site evaluations for a created system.

In restoring wetlands, the most important factor is likely to be understanding the hydrology since drainage and/or filling are most often the degradation factors. In that case, similar site investigations should be undertaken as with created wetlands projects before seemingly simple but possibly incorrect solutions — plugging the drains — are implemented. Destroying drain pipes or tile and plugging ditches may be successful in isolated wetlands, but if the subject system is part of a complex, most of which has been drained, interrupting drainage devices may not restore previous water levels since neighboring wetlands may have supplied water to the site through groundwater interconnections.

Excepting soils and climate, information on most factors identified below will be just as critical in a restoration project as a creation project. Presumably, the climate has not drastically changed though local modifications could have resulted from regional development and presumably the soils have not been completely altered. However, complete drainage, intensive agriculture, and possibly heavy erosion over a long period of time could have been more destructive than filling and paving for a parking lot. At least in the latter case, the original soil, and seedbank, are probably beneath the fill and asphalt and may start recovery after removal of paving and fill material and reflooding the site.

GENERAL CONSIDERATIONS

The extent of site evaluation will vary with type and magnitude of the proposed wetlands, but many important components of siting are common to virtually all development or land alteration projects. Wetlands planners must concern themselves with the same regulatory, land use, topographic, geologic, and hydrologic issues that face housing or industrial developers.

Types of information that must be assembled and analyzed include

1. land ownership, use, and availability
2. topography and geology
3. hydrology
4. soils
5. climate and weather
6. biology
7. regulations

Land Use and Availability

One of most important categories to investigate is land ownership, use, and availability. This includes title searches on the proposed site and adjacent lands to identify ownership, easements, rights-of-way, covenants, water rights, liens, and other encumbrances. Though importance varies with different regions, subsurface (mineral) rights and legal claims need to be determined as well as surface ownership since activation of a 100-year-old deep-mining claim might disrupt site hydrology and a new surface mine could substantially alter topography of the new wetlands. In western regions, an outstanding water right may nullify a planned water source, eliminating further consideration of that site.

Flowage easements on lands adjacent to reservoirs and rivers usually grant the right to inundate the land and may preclude deposition of fill that would reduce reservoir storage capacity. Watershed covenants may prohibit any activity deemed detrimental to the quantity and quality of runoff waters. Intricate webs of utility (water, sewer, gas, and electric) easements and rights-of-way crisscross the country, touching a large proportion of the landscape. Depending on the type of utility, wetlands construction and operation may or may not be compatible with dedicated uses. For example, proposing to impound waters above a gas pipeline would not be favorably received since leak detection and repair would be impaired. Conversely, building small ponds below a large transmission line, without impacting tower footings, might be acceptable to the utility but could be unwise since attracting large birds to the complex of wires might increase wire-strike mortality.

Ownership and availability (i.e., willing sellers or lengthy and costly legal maneuvers) may be the most crucial factor in selecting a site. Historical as well as present land use may favor or eliminate a site because of acquisition or construction costs. Adjacent land use is also important since it could detrimentally impact functioning of a wetlands or the wetlands may have detrimental impacts on current or planned uses of neighboring lands. Intensive agriculture with sediment and chemical-laden runoff adjacent to the site could impair wetlands functions and foreshorten its useful life. Conversely, increased humidity and bird populations near a major airport could endanger the health and safety of airline travelers; and building a "swamp" near an expensive housing development or office complex currently is not thought to improve property values, though public attitudes are changing. In addition, identifying undesirable land uses, toxic dumps, etc. could alert planners to potential costs of complex litigation and cleanup liability encumbering a potential site.

Developers should plan to acquire surface rights through fee purchase or long-term leases and easements. Fee simple acquisition is preferable to easements since small differences in cost may easily be overwhelmed by future complications. Unfortunately, easement rights are worth little when negotiating to obtain outstanding rights, that is, acquiring fee simple at a later date. In most cases, easements will cost nearly as much as fee simple and later costs for remaining rights will be about the same, resulting in purchasing the land twice.

Depending on the projected operating period and nature of other outstanding rights, mineral rights, water rights, and easements may also need to be obtained. Utility rights-of-way generally grant the right of passage, guarantee access for maintenance, and forbid incompatible uses. In some cases, discussing and negotiating planned uses may develop mutually agreeable solutions. In others, acquisition of easements and costs for relocating pipes or lines may need to be included in project cost estimates or an alternate site evaluated. Within reasonable limits, free title to project lands is generally worth the initial price in order to reduce future conflicts and save additional purchase costs at a later date, possibly in an adversarial environment.

Planners should also examine access for site inventory personnel, construction equipment, operating staff and equipment, and for utilities (electricity, phone, etc.). If public access is not available, corridors will need to be acquired in fee simple for people and equipment, though easements may be used for utilities. Distance and costs of extending utility pipes or lines will also be needed in preparing project cost estimates.

Topography and Geology

Site topography — elevation differences and spatial relationships — influences construction costs, erosion potential, drainage patterns, access, and overall feasibility. Shallow wetlands require relatively flat lands but few sites of any size are level. Since earth moving to create level to very gently sloping terrain is second only to land costs in most projects, accurate, detailed contour mapping is essential to project design and planning. During site comparisons, published maps at 5-foot contour intervals are adequate, but mapping to 1-foot contour intervals or less will be needed for final design.

Necessary evaluations of site geology include nature and depth to bedrock, potential construction materials, and other subsurface characteristics. Sites with shallow bedrock can dramatically increase construction costs because large quantities of rock may need to be broken up, removed, and replaced with imported fill. Costs and construction feasibility vary with type of bedrock because some types may be broken with equipment, whereas hard, continuous formations will require blasting. Erosion of exposed or near-surface bedrock creates parent material for soil formation and soil properties in the watershed can often be estimated from knowing the type and composition of underlying rock formations.

Underground limestone formations are often revealed by characteristic surface features typical of karst geology, the kettle and dome patterns, sinkholes, and caves or caverns. Most were formed by dissolution, often subsurface, of limestone and, in moderate to wet climates, this process is on-going. If the proposed site has limestone strata or karst topography, detailed seismographic surveys will be needed. Sudden appearance of an open sinkhole or land subsidence in a new wetland renders water control difficult, if not impossible, and is costly to repair. In arid limestone regions, crevices or channels may have formed under previously wet climates that could similarly jeopardize project goals. Soils formed from limestone parent materials also tend to be highly permeable and difficult to modify to seal pond bottoms.

Most of our land has been explored in more or less detail and much has been actively mined or drilled. Failing to identify old mine shafts, tunnels, air shafts, and bore holes could similarly cause sudden drainage of the new system, while uncapping a "dryhole" in an oil field might suddenly introduce volumes of brackish water or undesirable gases.

Soils

Soils and parent materials should be evaluated for class and composition, distribution, and depth. Parent or subsurface materials form dike and dam foundation and will be used as construction materials, and topsoil becomes the substrate for plant growth. Important soil attributes include proportions of silt, sand, clay, gravel and organic material, texture and particle size, permeability and drainage potential, erodibility, and chemistry. Evaluations should include potential borrow areas adjacent to the site if fill, dam coring, or bottom lining materials will be needed.

Hydrology

Evaluating site hydrology includes understanding surface and groundwater location, quantity, and quality, along with surface and subsurface flow patterns, connections, and seasonal changes. Obviously, the amount and type of surface water in the drainage basin influence the size, nature, and operation of the proposed wetlands; but subsurface waters and direct connections with surface waters may have considerable impact in some circumstances. Maintaining necessary water levels will be difficult with inadequate runoff and alternate supplies may need to be tapped, at least during dry periods. High sediment loads and agricultural, industrial, or mining pollutants in potential water supplies will need to be analyzed and compensated for if the new system is to provide safe haven for wildlife and recreation opportunities for area citizens. Conversely, clean water and poor soils could require fertilizing after planting to improve nutrient supplies for biological growth. If the wetlands are planned as a treatment system, volumes of flow and type and concentrations of contaminants should be well documented for at least 1 wet and 1 dry year to provide parameters for system design.

Flooding and accelerated erosion potentials within the basin must be determined so that appropriate protective measures can be designed to accommodate expected runoff. Wetlands near streams or rivers will be susceptible to flood damage to vegetation and physical structures unless protective measures are included in the design. Since permanent streams are often connected with or at the same levels as groundwater, sites near streams may or may not be suitable, depending on the use of the new wetlands. All streams are not "gaining" streams; some are "losing" and depth to groundwater increases rapidly with distance from the stream (see Figure 2). Locating a wetland in the valley of a losing stream may become an exercise in frustration over inability to seal the bottom. Similarly, depending on groundwater for water supplies may or may not be appropriate since little control over the basic management mechanism (hydrology) will be possible and falling groundwater levels could jeopardize the continuance of the wetlands.

Springs, sinkholes, and other significant connections between surface and subsurface waters must be identified and considered during design, lest construction activities disrupt normal flows within or downstream of the site. Emerging waters or near-surface water may impact dike and water control locations, hinder construction activities, and influence system operations. It may be necessary to design temporary

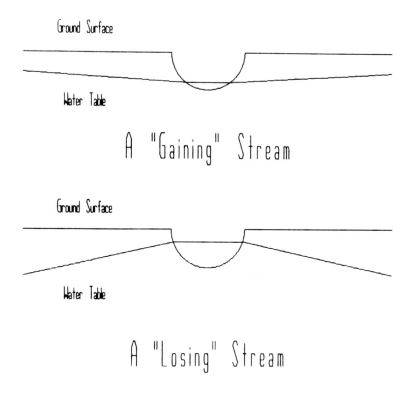

Ground Surface

Water Table

A "Gaining" Stream

Ground Surface

Water Table

A "Losing" Stream

Figure 2.   In wet climates, stream water levels often depict groundwater elevations, but in much of the West, streams carry water down from the mountains into arid plains where water tables may be much deeper than stream water levels.

diversions during construction or permanent structures to include or exclude these sources from future pools. Though pumping from wells will drastically increase operating costs, subsurface waters may provide supplementary sources in critical drought periods.

Groundwater descriptions should include depth, quality, isolated or perched water tables and flow patterns. Depending on surface water supplies and type of wetlands, a groundwater discharge or recharge area might be advantageous. Discharges through springs or seeps could supplement surface waters during operation, though hampering construction. Sites with dry streambeds, fissured metamorphic or igneous rock, and karst geology should be avoided if the wetlands are to be used for wastewater treatment. Known recharge areas of any type could be difficult to seal and water losses might be intolerable, especially during dry periods. Of course, if recharge is the design function, then the wetlands should be sited over a recharge area.

Locations and type of existing and planned use of surface and groundwater at potential sites should be determined to avoid interrupting current activities or to estimate acquisition or mitigation costs. Potential adverse impacts from other users, as well as to other users, should be estimated during site evaluation. Existing impoundments, especially large reservoirs, in some regions cause substantial increases in groundwater supplies, while ditching and draining cause reductions in most areas. Both can provide indications of groundwater status.

Climate and Weather

Weather patterns and climatic regimes restrict choices of wetlands types, dictate biological components, and influence operating procedures. Daily and seasonal precipitation patterns (including frequency, intensity, and duration of storm events) are determinant factors in amounts and timing of runoff crucial to most wetlands. Temperatures can also have important effects on amounts and pattern of runoff. Daily and seasonal air temperatures and, indirectly, water temperatures affect basic chemical and biochemical processes that are the basis of all life forms. Prevailing wind velocity (direction and speed) affects orientation of ponds and perimeter planning during design because of wave action impacts to vegetation and dikes, and water loss from evapotranspiration during operation. Regional and local relative humidity and incident solar radiation influence and interact with precipitation and temperature, thus controlling many biological and physical processes.

Climatic data are from standard observations at airports, NOAA weather recording stations, Coast Guard facilities, and a few other specific sites. While it provides a broad overview of past and expected weather in the region, numerous topographic and biological factors influence generation, location, and diversity of weather patterns affecting a specific site. Generally, the more distant and the more varied the topography between the recording station and the candidate site, the more variation in actual weather is to be expected. Climatological Data, an historical summary for various areas of the country is available from the U.S. National Oceanic and Atmospheric Administration, Asheville, NC.

Hills or ridges, mountain ranges and valleys, large water bodies, upsloping though relatively level terrain, presence and type of vegetation, and north or south-facing slopes impact precipitation, humidity, solar radiation, and wind currents, creating microclimates that might be quite different from regional patterns (see Figure 3). North-facing slopes often have growing conditions typical of more northerly regions, while the opposite is true on the other side of the ridge. Lake effect precipitation and temperatures are well known to residents of western New York, as are sea and land breezes to coastal inhabitants. Up-sloping terrain perpendicular to prevailing winds causes increased precipitation on the windward side and reduced rainfall on the downwind side. The striking differences are evident to a casual observer crossing the divide in the Cascades and other western ranges. Mountain valleys often generate daily wind patterns with up-slope currents during the day and down-slope, cold air drainage at night. Presence and type of vegetative cover reduces wind action and air temperatures, and increases relative humidity in local areas. Patchwork vegetative cover modifying solar heating results in thermals — rising columns of air over bare ground and descending columns over dense vegetation. Since thermals are the genesis and progenitors of thunderstorms, vegetation patterns also influence local precipitation patterns.

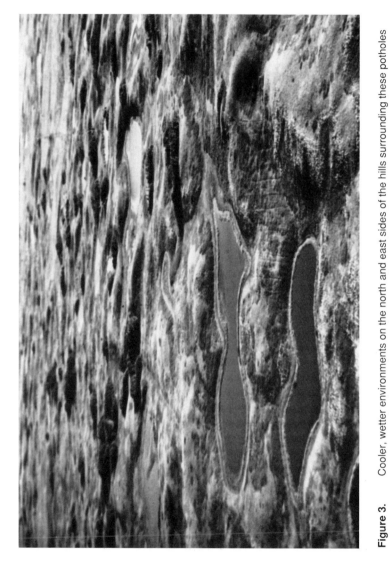

Figure 3.   Cooler, wetter environments on the north and east sides of the hills surrounding these potholes support woody plants, whereas the warm, dry southern slopes are grass covered.

Length of growing season is a generalized parameter that manifests the effects of many climatological variables influencing types of plants or animals that can survive on the site. It obviously will affect selection of plant species to be used, as well as operating constraints for some types of wetlands. However, general values represent regional conditions that may need interpolation for describing a specific area or site.

Biology

Biological factors in site evaluation fall into four categories: (1) animals or plants that would be damaged or detrimentally impacted by construction of the wetlands; (2) those that would be detrimentally impacted by project operation; (3) those that may detrimentally impact the new wetlands; and (4) those that facilitate data collection on other aspects of the site.

Creating a wetlands on a terrestrial site will obviously be detrimental to the previous inhabitants and, if any species of plant or animal has unusual aesthetic, uniqueness or commercial value, that loss must be factored into site evaluation and selection. Earth moving to restore an impacted wetland is more likely to impact a unique species since so many wetlands species have become rare or isolated because of past wetlands drainage.

New or additional wetlands habitats may alter local or even continental distribution patterns for wetlands and some terrestrial species. For example, creation of new habitats during the last 40 to 50 years in mid-latitude states — New York, Illinois, Ohio, Kansas, etc. — has dramatically altered migration and wintering patterns of continental waterfowl populations with few wintering in previously used habitats in southern states. Changes in local distribution patterns may bring various species into conflict with other land uses or human activities. In general, restoration or creation has a long way to go before it can replace the original wetlands resource base, and most new systems will have overall positive impacts to wetlands wildlife.

Presence or proximity to existing wetlands may be important for planting materials or could be setting the stage for invasion of undesirable weeds. Creating a new wetlands near an existing system that is heavily infested with purple loosestrife (*Lythrum salicaria*) will increase operating costs of the new system.

Presence of plants or animals is perhaps the best indicator of site suitability since they reflect the culmination of present conditions of soil, water, climate, and land use. Although information on hydrology, soils, climate, etc. is important, the significant features of the new wetlands are the biotic components. Not surprisingly, they often have similar requirements to survive and grow on the site as do the present occupants. Length of growing season, availability of soil moisture, permeability of soils, fertility of soils, and past and present uses can all be interpreted from biological inventories of native or cultivated species. For example, a soybean field in the Mississippi Delta quite likely occupies a previous bottomland hardwoods site with moderately fertile, hydric soils (see Figure 4).

If only sedges and wet grasses are found in a high mountain valley, attempting to create a cattail or bulrush marsh is likely to fail because of the short growing season.

Figure 4.   Agricultural fields occupying previous wetlands sites in eastern Arkansas would quickly revert to early successional wetlands if drainage ditches were plugged.

Bladderwort and pitcherplants indicate reduced nitrogen, chironomids indicate low dissolved oxygen or other pollutants, lichens reflect air quality, blue gramma grass suggests native prairie, and of course wetlands species suggest hydric soils and at least remnant wetlands systems. *Scirpus fluviatilis* and *Ruppia maritima* are found in brackish waters unfit for establishing *Nuphar* or *Pontederia,* and cottonwood, willow, or salt cedar do not indicate a good site for cypress or gum. Birds and mammals tend to be less useful because of their mobility, but amphibians, insects, and other invertebrates (especially aquatic types) may be valuable indicators.

Regulations

Laws and regulations that may impact project construction and/or operation need to be considered as soon as potential sites have been identified. Though most vary from state to state, many fit within a federal framework and others are familiar to federal regulatory agencies that will be contacted for information on federal laws and regulations. Bear in mind that even though the proposal is to create or restore a wetlands and the proponents may be concerned environmentalists, the same laws and regulations governing dredge and fill or other degradation activities must be complied with by those proposing to build wetlands. Some wetlands proponents have the impression that creating or restoring wetlands is an environmentally sound activity and therefore immune from environmental regulations. Fortunately, this is not true. Though most will be impatient with the regulatory review process and some may be affronted or at a loss to prepare an adequate justification for their proposal, wetlands developers are not infallible and they are just as likely to alter flood patterns, impact an endangered species, or damage a cultural site as someone proposing a housing development or marina.

The most important regulations cover earth moving (dredge and fill) in rivers, streams, and wetlands, protection of endangered species habitats, protection of wetlands and floodplain management, and preservation of cultural and historic resources. Specific federal laws include:

Clean Water Act
Executive Order 11988 — Floodplain Management
Executive Order 11990 — Protection of Wetlands
National Environmental Policy Act
Fish and Wildlife Coordination Act
Endangered Species Act
National Historic Preservation Act
Clean Air Act

State or local laws and requirements are often a subset of the federal frameworks, though in some instances, local regulations may be considerably more stringent. In addition, most states have regulations governing impounded waters over some minimum size that should be examined for pertinence. The state clearing house generally coordinates review of projects by various state agencies and can provide information on applicability of state regulations.

Low-lying or poorly drained lands are often optimal for creating or restoring wetlands since the site may have supported a wetlands system in the past. However, construction activities in areas with hydric soils will likely entail dredge and fill actions that are covered by Section 404 of the Clean Water Act and will require appropriate reviews and permits. If a favored site is wet, developers should initiate coordination with the U.S. Army Corps of Engineers at an early stage. Executive Orders 11988 (Floodplain Management) and 11990 (Protection of Wetlands) are applied by Federal Agencies during review and permitting processes under Section 404, Section 10 of the Rivers and Harbors Act, or other federal legislation. Neither Executive Order has legal standing (other than for federal agencies) on its own, but both are guidelines for the interpretation and application of federal laws. Consequently, either or both could be germane through the 404 process if the chosen site has or had wetlands characteristics or is in a floodplain.

Since many threatened or endangered species are wetland types, any restoration project could easily fall under the purview of the Endangered Species Act. Alternatively, changing terrestrial habitats to wetlands in a creation project could detrimentally impact a rare terrestrial species. Most state offices have a Natural Heritage Program sponsored or initiated by The Nature Conservancy, that will have detailed distribution maps for federally listed threatened or endangered species. Their database often also includes information on species under consideration for federal listing and on species that may have protection under state threatened and endangered species laws or regulations. The nearest office of the U.S. Fish and Wildlife Service has similar information and has statutory responsibility for administering the Act. In most cases, early contact will eliminate any further coordination or action, but in some instances, detailed surveys may be necessary.

Stream bottoms and river valleys were favored habitations for past human cultures probably because of the same attributes that make them preferred sites for wetlands creation. Unfortunately, much of our cultural and historical resource has been lost to past development, but recent federal and state laws and regulations provide fairly comprehensive protection today. The state historic preservation officer can assist project planners in identifying potential cultural sites in the area and advise on inspections or surveys as well as protective measures that may need to be employed.

SPECIFIC CONSIDERATIONS

Locating wetlands to provide recreational or commercial marketing benefits has less constraints than more restricted uses. A constructed wetland (wastewater treatment) site is often predetermined by the location of the wastewater source, that is, an industrial site, a municipal treatment plant, or an acid drainage seep. Since the wastewater source can only be relocated with costly piping or pumping, siting the treatment system is usually limited to evaluating a limited geographic area, which may or may not be an optimal site.

Site evaluation for constructed wetlands should include considerably more data collection on adjacent or downstream lands and evaluation of potential impacts from the constructed wetlands. Detailed chemical descriptions of receiving waters and estimates for expected discharge quality should be obtained to evaluate and monitor potential impacts. An early meeting with adjacent landowners to thoroughly explain proposed activities, expected impacts, and project goals is highly recommended.

Planners of flood storage/buffering wetlands, ground water recharge systems, or education facilities are likely to have similar but not as exacting constraints. Hydrologic buffering wetlands must be located in the floodplain, but there may be ample latitude within the valley. Information on nearby and downstream lands will be important in evaluating potential impacts to these areas during operation. Groundwater recharge systems may also have more latitude than wastewater treatment wetlands, depending upon the location and extent of the pervious layers. Many recreational systems offer considerable siting flexibility, but educational systems generally need to be near or on school facilities and site selection will be limited.

Data Collection

Much of the information needed to evaluate prospective sites is available from local, state, and federal agencies. The first step is to contact these offices for their information on your site(s). After completing this phase, additional steps may be necessary as outlined below:

1. compilation of available information — office
2. interpretation and evaluation of available information — office
3. ground or aerial inspection of site(s)
4. preliminary data collection on site(s): surface water quantity and quality, subsurface water parameters from available wells, soil classifications and profiles, subsurface formations, and biological communities
5. extensive or detailed surveys on the above categories as needed
6. comparative evaluation of data, estimating feasibility and costs for development and operation at candidate site(s), and preparation of preliminary budget

Data collection efforts for site evaluations must be reasonably adjusted to the magnitude and complexity of the project. Years of effort and thousands of dollars for siting to build a 10-acre system could rarely be justified. Conversely, a low-budget, cursory review for a 1000-acre or multimillion dollar wetlands may result in inordinate construction or operating costs when serious problems are discovered later; or the system simply may not function as planned. Most of the information is likely to be available for the asking and should be examined regardless of project size. However, detailed site surveys can be costly and should not be initiated unless necessary to fill in crucial gaps in available information.

Information management systems should be considered in the earliest stages of data assembly. If planners have some form of geographic information system (GIS) support, data should be entered into this system as it is assembled even though data entry may seem costly. The benefits of using a GIS for displaying information during comparative evaluations of sites far outweighs the cost. With the proliferation of personal computers, a variety of evaluation and management simulating programs have been developed that are useful in siting, as well as modeling development/

management options. One of the oldest is HEP (Habitat Evaluation Procedures) developed by the U.S. Fish and Wildlife Service. Contact the Division of Ecological Services in Washington, D.C. for information on availability of software and manuals. Two others developed by the U.S. Fish and Wildlife Service, National Ecology Center in Ft. Collins, CO are micro-HSI (Habitat Suitability Index) and HMEM (Habitat Management Evaluation Method). A method for assessment of wetlands functional values, Wetland Evaluation Technique (WET), developed by the Federal Highway Administration has been adapted for PC use by the U.S. Corps of Engineers, Waterways Experiment Station, Vicksburg, MS. The Fish and Wildlife Service has recently developed a computerized ranking system to facilitate land acquisition prioritizing. Contact the regional office of the U.S.F&WS for information on LAPS (Land Acquisition Priority System). Finally, WETLANDS is an electronic data base of state wetland protection programs and contacts developed by the Council of State Governments, Box 11910, Lexington, KY 40578.

Though complexity and cost of data entry into a data management system may seem high, the site selection process will be substantially improved. In addition, most organizations store current information in an electronic form and many have or are in the process of transposing historical records. Inquire about the availability of the data on diskette as you contact each organization. Lastly, creating the information baseline as a management tool for operation of the new wetlands system at the chosen site may be just as important as data management in the siting process.

Sources and Examination of Available Information

Topographic, geologic, geologic hazard, mineral deposit/exploration, and hydrologic maps and text available from the U.S. Geological Survey (U.S.G.S.) should be examined for relief, elevations, presence and type of waterways, sinkholes or other subsidence indicators, cultural facilities (buildings, roads, etc.), utility lines, mines and wells, abandoned mine shafts, springs, and location and nature of groundwater. Degree of detail varies considerably with the type of information and the area of interest, which is the reason that planners should contact the local U.S.G.S. office for advice rather than simply order the maps or descriptions from the central office. Mapped contours on the common 15-min quad sheet at 1:125,000 scale may not be adequate in mountainous terrain since contour intervals could be 20 feet or more depending on the amount of relief on an individual map.

Much of the U.S. has been mapped for the U.S. Fish and Wildlife Service's National Welands Inventory (NWI). NWI maps depict location, extent, and type of wetlands in overlay or composite formats at scales of 1:24,000 to 1:250,000. They are available from a number of state or regional offices of geological survey, wetlands, and map distribution centers or from the Earth Science Information Center, U.S. Geological Survey, 507 National Center, Reston, VA 22092; 800-USA-MAPS, (703) 648-6045. The NWI has also digitized slightly over 11% of the NWI database and information on data products is available — (813) 893-3873. In addition, MicroImages, Inc. (201 North 8th Street, Suite 15, Lincoln, NE 68508; (402) 477-9554) has digitized maps with wetlands and typical cartographic features available for some areas.

Soil surveys, aerial photography, and a variety of information on growing seasons, plant requirements, fertilizer recommendations, and pond construction can be obtained from the district office of the SCS or ASCS. Specialists in these offices are experienced in almost any aspect of farming, much of which pertains to building and operating a wetlands system. Frequently, they are also locally knowledgable about groundwater conditions and, to some extent, subsurface conditions including bedrock.

Older, perhaps sequential, aerial photography can often be obtained from the ASCS, the U.S. Geological Survey, a local mapping agency such as the Tennessee Valley Authority in the Tennessee Valley region, various other federal and state agencies, or an aerial photography service. Try to determine if any agency or organization may have studied the region for some purpose and then ask if they kept the old prints or negatives. If sequential photography can be obtained, it will prove invaluable in determining site history. Factors such as biological communities, frequency and extent of flooding, drainage, erosion, sedimentation, land use, abandoned or reclaimed mines, buried landfills, waste disposal sites, pipelines, and property ownership (field boundaries from fence or tree lines) can often be detected and/or followed in a sequence of aerial photos.

Careful interpretation can even identify cultural and archaeological resources such as evidence of ephemeral use of an area by historical inhabitants. For example, the Plains Indians collected small rocks to weigh down the fringe of their tipis and then rolled the rocks off when the camp was moved. Even today those clusters of rock rings marking an old camp are clearly seen in aerial photos or low altitude flights over remnants of the native short-grass prairie. Not only do these rock patterns reveal cultural resources, but the fact that circular patterns are still present also indicates the land has continuously supported native prairie. Many other parcels may have prairie today but were cultivated at some time in the past and the rock rings were either disrupted or rocks were deliberately cleared to facilitate use of farm machinery.

Large-scale surface geological formations are much more evident from aerial photos than from ground inspections and even subsurface formations may have surface representations. For example, boundaries of cedar glades, a unique terrestrial system in Kentucky, Tennessee, and Alabama caused by near surface limestone formations, are revealed in springtime aerial photos or overflights by the distinctive greenish foliage of prairie clover (*Petalostemon gattingeri*).

Available imagery may go well beyond the expected black and white or color photography. Depending on the area, organizations with different interests and responsibilities may have obtained IR, UV, vertical or side-looking radar imagery, magnitometer surveys, and a variety of other specialized data types. Most can be very useful, but some may need to be interpreted by specialists. For example, plant stress from poor soils, inadequate water, disease, or insect infestation is easily detected by the pale or light reddish color in IR photos because healthy vegetation tends to have dark shades of red. Different species also cause different shades, but considerable experience may be needed to correctly interpret the variations. Understanding magnitometer survey data or side-looking radar images is a bit more complicated. On the other hand, dominant species composition of a forest is easily determined from color aerial photos made during the fall color season and wetlands maps (delineation and classification) for the NWI are prepared by photo interpretive specialists from high-altitude photography with occasional ground truthing (see Figure 5).

Flood hazard maps available from Federal Emergency Management Agency depict expected flood frequency in terms of area inundated by 50-year, 100-year, or 500-year floods and can be useful in identifying flood ways, flood plains, and flood-proofing requirements.

All in all, the proportion of information available for most sites from maps, aerial photos, surveys, and textual descriptions is likely to be much greater than will be obtained from typical on-site surveys. Only if planners initiate extensive, detailed and expensive surveys of all factors important in siting are they likely to obtain an equivalent amount of useful siting information. Much of the available information is free or provided at nominal cost and can be reviewed, categorized, and evaluated with only the cost for the evaluator. Consequently, the majority of siting effort should be placed on office collection and interpretation of available information. Other than a low altitude overflight, new aerial photography or a walk over, on-the-ground surveys should only be planned and initiated for major information gaps in critical siting factors.

After office comparisons have been completed, low altitude (500 to 2000 m above ground level) overflights will expedite verification of assembled data for alternative sites, especially if the sites are widely separated. Flights should be scheduled in early morning hours on sunny, calm days to avoid thermal turbulence and because low-level sun angles and subsequent shadows improve topographic perspectives. Depending on factors of interest, presence or absence of tree leaves and other foliage may obstruct or enhance site evaluations and flights should be planned accordingly. Choice of altitude is a compromise between scale and fine detail and time over a specific area since relative speed increases with lower altitudes reducing the available time to study specific details. Generally, a high-altitude pass followed by one or more overflights at lower altitudes is the best combination.

Other types of useful information include county and regional maps, drilling records from water, oil, and gas wells, stream water and biological surveys, permit records from mineral or petroleum explorations and highway construction, environmental impact statements, and zoning regulations. Even patterns and names on road maps can be useful. Straight roads in England are striking evidence of old Roman roads, in contrast with typical serpentine roads of other periods. Turnpikes, tollways, and ferry roads are often names that reveal historical routes that may have associated cultural resources as do names including fort, mill, or mill pond. Additional information sources include government agencies, universities, local builders, surveyors and contractors, utility companies, and naturalist or sporting clubs and societies.

After completing assembly and evaluation of data collected from the above largely remote sensing methods, permission to inspect the site should be obtained and a careful walkover carried out with specialists as needed. Most likely, information will be scanty or absent on one or more factors that can be supplemented by a simple field examination by an experienced professional. Trips to the site are also opportunities to determine access and discuss site history with local residents, either owners or neighbors, as well as verifying other information collected to date.

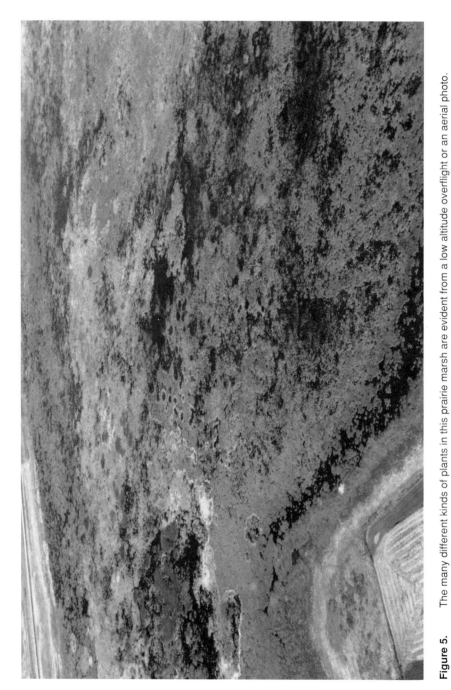

Figure 5.   The many different kinds of plants in this prairie marsh are evident from a low altitude overflight or an aerial photo.

Most likely, detailed topographic mapping will be necessary at some stage in the process since planning will require 1 foot or smaller contour information. Other on-site surveys that may be useful include auguring, test pits, percolating and compaction testing, along with limited surveys of vegetation, soils, hydrology, cultural resources, and geologic formations such as rock outcrops. Current and expected use of the site and neighboring lands may often be identified during site inspections and interviews with residents.

During these contacts, project goals should be carefully and fully explained to gauge public attitudes and reactions, or to modify attitudes if necessary and feasible, before major commitments have been made. Though wetlands are virtuous in some circles today, remnants of the "dark and dismal swamp" persist, as do unpleasant experiences with some aspects of wetlands. Concerns over snakes, mosquitos, odors, depressed property values, aesthetics, or crop depredation must be addressed frankly and patiently even though many derive from misinformation.

At this point, planners should have enough information so that significant deficiencies, if any, are obvious. Detailed contour mapping, perhaps laboratory analysis of soils, test pits, or shallow drilling for groundwater, bedrock information, and borrow materials are the most likely surveys that will need to be conducted. However, some sites may potentially have a listed rare or endangered species, significant archaeological or historic resources, subterranean caverns, or unusual groundwater distributions, and comprehensive surveys must be initiated. Generally, the responsible regulatory agency will be able to informally advise on local or regional expertise that should be obtained if all other factors favor that site.

Extent and detail of these surveys, as with other means of information collection, must be representative of complexity and magnitude of the project. Minimally, the siting process will have assembled the information needed to identify hazards to be avoided and to develop plans and cost estimates for project construction and operation during the planning stage. If additional information seems necessary, astute planners will obtain it since an overly conservative design may be much more costly to build and operate than the apparent cost savings from limited site investigations.

DESIGN SELECTION

Perhaps the most enjoyable part of this process is designing the new wetlands. This is the time to release the creativity of your imagination, to translate ideas into drawings, and to revise ideas and drawings to accomplish your objectives, all without exceeding available funds. But once a site is selected, the properties of the site may accommodate your wishes or they may place serious constraints on your innovative concepts.

By determining the functional values sought from the new wetlands, you have already narrowed the choices of wetlands types. Though certain kinds are more suitable for certain objectives, the site you have selected or are forced to use may further restrict the usable options. Previously unbridled design innovations may need to be tempered or even completely revised after a careful review of the properties of your site. Or constraints imposed by site properties may preclude building and operating the type of wetlands that will provide the functional values listed in your objectives. Two characteristics, water supply and topography, are perhaps the most important factors to consider since either may eliminate some designs. The need for an adequate water supply is obvious. Topography may be less evident, but it is no less important. Relief or the differences in elevations, and spatial relationships, the locations and proximity of high and low areas, are important determinants of the potential size, shape, and depth of the new system. Of course, seasonal fluctuations in runoff or possible alternative supplies and the seasonal needs of various wetlands types, soil and parent material type and condition, property boundaries and adjacent land uses, and types of natural wetlands in the region must also be evaluated before selecting a specific design.

Examining natural wetlands is important because the created wetlands must closely imitate natural systems adapted to that region if it is to succeed without excessive operating and maintenance costs. Since we hope to create a system that will be largely self-maintaining, we are not likely to succeed if we attempt to build a wetlands type that is adapted to entirely different climatic, hydrologic, and edaphic conditions. Regardless of the functional benefit expected from the new wetlands (the objectives), the new system should mimic natural wetlands as closely as possible. To attempt otherwise may be frustrating and expensive.

Figure 1.   Mallard and pintail ducks use many different types of wetlands in different regions of the country for breeding, migrating, or wintering.

Available Water Supply

Adequate supplies of water at the appropriate time of year are crucial to sustaining a wet system. Do normal precipitation in the region and runoff from the specific watershed match the size and shape of the new system? How large a watershed is needed to supply adequate runoff for the preferred type or size or shape? Must the surface area be reduced to cut down on water losses, or should another watershed be investigated? Answers to most of these questions are developed from examining precipitation, evaporation, and evapo-transpiration values for the region and a particular site. Understanding these factors in conjunction with variables influencing amounts of runoff provides valuable insight into the feasibility of creating a functional, self-maintaining wetlands. Not surprisingly, wetlands in arid regions need runoff from a larger area of watershed to maintain adequate water levels during the driest part of the year and, even then, may lose depth or dry out during drought years. Conversely, a small wetlands fed by a large watershed in a wet climate will not lack for water but may require elaborate and costly water control structures and emergency spillways for flood protection.

The first step is obtaining average monthly values for precipitation and evaporation losses in the district and then estimating evapo-transpiration losses from the types of wetlands under consideration. Precipitation and evaporation losses, commonly combined as the precipitation/evaporation ratio (P/E), may be obtained from the nearest office of the U.S. Weather Bureau, the Soil Conservation Service and many extension agents, or from climatological data compiled by the National Climatological Center in Asheville, NC. Precipitation values are derived from long-term averages of actual precipitation records. Evaporation values may also be developed from historical observations or they may be computations of the quantity of water expected to evaporate from an open water surface under prevailing conditions of air temperature, relative humidity, and wind speed over a period of time. In either case, they are commonly referred to as lake or "pan evaporation." In simplest terms, the annual or monthly P/E ratio predicts the amount of water that would remain in an open-topped shallow container, a pan, after precipitation inputs are reduced by evaporation losses during the specified interval.

Though this ratio is valuable information, it may need slight modification since a wetlands may lose water to percolation into underlying soils and because vegetation is not present in deep-water lakes or the abstract pan. Plants complicate the analysis by transpiring — losing water from cell surfaces directly exposed to the atmosphere to exchange oxygen, carbon dioxide, and other gases. Conversely, plants obstruct air flow near the water's surface. In windless regions or under still conditions, the atmospheric layer nearest the water surface becomes saturated and little additional evaporation occurs. But under windy conditions and turbulent flow, the boundary air layer is replaced continuously. If the relative humidity is high, evaporation rates may not increase, as is often the case on a still night. On a hot, dry day, continuous renewal of the saturated boundary layer with dry air may dramatically increase evaporative losses. Clearly, anything that obstructs or influences air flow patterns near the water's surface will influence evaporation rates and the effect will be greater in hot, dry weather than in cool, wet periods. Plants, both herbaceous and woody species, along the edges or within the wetlands have variable impacts on air flow patterns depending on the density, height, and areal coverage of the stand.

If the selected wetlands type is primarily shallow open water, standard P/E ratios fairly approximate water balances. However, in large, dense stands of tall plants, cattail, bulrush, reeds, or woody species, transpiration losses from photosynthetically active plants become significant and the important parameter becomes the P/E ratio. Though transpiration losses were once thought to be additive to pan evaporation losses, recent measurements have shown that combined loss from evaporation and transpiration in a densely growing emergent species is approximately 80% of pan evaporation loss in a specific region. Apparently, lower wind speed and exposed water surface coupled with high relative humidity within an emergent stand substantially reduces direct evaporative loss from the water's surface and, even with the addition of transpiration losses, the combined total is lower than from open water surfaces.

Obviously, a P/E ratio can be computed for the entire year, but even a cursory review of factors contributing to this expression suggests that monthly values should be examined. This is especially important when designing a wetlands in drier regions. P/E values change, sometimes dramatically, during different seasons and evaporation losses during the driest part of the year may far outstrip precipitation inputs imperiling water levels in wetlands. Hot, dry months with low P/E ratios may cause dramatic evapo-transpiration losses from wetlands without compensating inputs from runoff. Cool, wet months with positive P/E values will likely support higher water levels because losses are reduced and runoff is increased. However, runoff extremes must then be considered in designing flood protection measures.

Northern climates further complicate the evaluation. High P/E ratios during cold months do not simply translate into increased runoff because of evaporative losses (snow sublimation) during the winter and differential runoff of meltwater in spring. Large amounts of runoff may be generated by snow melting over frozen ground but much potential runoff is lost to infiltration when snow melts over unfrozen ground. Neither condition is unusual and both may occur in different parts of the watershed, depending largely on air temperatures and snow cover. Extreme cold prior to the accumulation of an insulating snow blanket may freeze soil moisture deep into the underlying parent materials. In spring, deeper layers melt later than surface layers and melt water runs offsite rather than percolating into the soil; runoff will be much lower following winters with substantial snow cover prior to extreme cold.

Annual P/E ratios are high — precipitation is greater than evaporation losses — east of the Mississippi River and low in the Great Plains, Intermountain West, and the Southwest Deserts. However, higher elevations (mountains) and coastal regions in both sections of the country receive more precipitation, cooler air temperatures reduce evaporative losses, and the P/E ratio is higher than surrounding areas. To some extent, forests are indicative of regional P/E ratios since most natural forests occur in regions with moderate to high P/E values. Riparian forests supported by high water tables and/or local microclimates (higher P/E) in western river valleys are evident but not unequivocal exceptions.

Generally, less than 5 acres of watershed is needed for each acre-foot of water in a farm pond east of the Mississippi River. Since shallow wetlands have greater surface area to volume ratios, additional watershed acreage will be needed. For example, a 20-acre wetlands with average depths of 0.5 feet would have 10 acre-feet of storage and nominally need a 50-acre watershed; but a conservative design would increase that by 50% for a total of 75 acres. At lower elevations in the West, watershed size for farm ponds ranges from 20 acres in the Plains to 140 acres in the Southwest. These are gross approximations for typical conditions. Actual runoff amounts are strongly affected by amount, intensity, and duration of rainfall as well as watershed soils, topography, and vegetation cover. Details on developing runoff estimates with precipitation descriptors and watershed characteristics are presented in the section on emergency spillway design.

However, these methods were developed to estimate watershed area needed to supply runoff that will sustain an open water pond. If the new wetlands is located in a humid region (P/E ratio >>1), is densely vegetated, has an impermeable liner, and has precise discharge control, somewhat less area will be needed because the *annual* evapo-transpiration rate from the wetlands will be approximately 80% of the evapo-ration rate from a comparable amount of open-water surface in a pond.

If sustenance water for the new wetlands is questionable, tall, dense emergents over most of the area is preferable to large, open waters if either type will meet project objectives. If not, consider including a zone of dense emergents or a shelterbelt on the upwind perimeter to reduce wind speed and evaporation losses from open-water areas. Remember, however, that in arid regions the roots of screening trees will grow towards water, perhaps into the wetlands. Commonly occurring willows and cottonwoods are poor choices for wind breaks since they do not conserve water as well as mesquite (*Prosopis*), acacia (*Acacia*), or locust (*Robinia*).

Irrigation water return flows; that is, water that has been used to irrigate cropfields or pasture may be a tempting source of normal operating or emergency supply, but should not be used under any circumstances in any region where the P/E ratio is less than 1. Farmers in a few areas, Georgia, Florida, etc., with humid climates have recently begun to use irrigation during the driest part of the year. In those areas (P/E ratio >1) return flows may be useable, after chemical analyses for fertilizers and pesticides, for temporary or emergency makeup water only. In dry climates, basically west of the Mississippi River, soil salt contents tend to be high and many areas have elevated concentrations of selenium. Typically, irrigated fields have a layer of salts 4 to 20 centimeters (cm) below the surface caused by long-term flooding for irrigation under high evaporation but low percolation conditions. Runoff from these fields contains high salt and often high selenium concentrations that will be deposited in the wetlands, bioaccumulated, and impact fish and wildlife. Marshes in a National Wildlife Refuge in California (Kesterson) are no longer useable, in fact dewatered, because of selenium concentrations caused by using irrigation water return flows for many years.

Site Topography

Site topography will guide, if not dictate, optimal locations for dikes or dams since closing off a narrows often requires less fill material than arbitrary or randomly placed dikes. Placement of dikes and relative elevations above the dikes determine the areal extent and depth of impounded waters and influences water level manipulation capability of the new wetlands. Size, shape, and depth of each pond in the new wetlands results from the ability of dikes and water controls to impound waters to elevations at or above upstream land surfaces. A broad, flat area will need only a low dike to flood a large area, whereas high dams may be needed in areas with high relief. The short, high dam may be less costly than a long, low dike, but the amount of shallow waters created and operational flexibility must be matched with project objectives. The proportion of shallow (5 to 20 cm) water depths needed for marsh plants or for bottomland hardwoods is large relative to the cost in the flat area, but small at the steep site. Conversely, rapid increases or decreases in water depths may be impractical without a large capacity or several water control structures in the flat area, while a large structure may quickly raise or lower water levels in a narrow valley. In the broad, flat area, rapid lowering of water levels to avoid stress on favored trees may not be possible in certain seasons without many control structures.

The deep linear system, protected from wind exposure to reduce evaporative loss, would be optimal for a groundwater recharge/infiltration system assuming it was underlain by highly permeable substrates.

Total dirt moving and grading requirements may often be reduced by carefully positioning dikes and dugouts. Construction grading will be a major expense and balancing cut and fill in nearby areas will reduce materials transport within the construction area or off-site. Fill material for dikes should be obtained by removing thin layers of earth from broad expanses rather than forming a deep borrow pit to maximize extent of shallow-water areas. Excess fill may be used to widen or increase dike height, form parking areas and observation turnouts, or to create nesting islands or hummocks in selected locations. Low gradient mounds later overtopped with shallow water can be used to form islands of shallow water species (bulrush or arrowhead) within selected deep-water areas, thereby increasing diversity of the system. Elevated ridges or mounds supporting terrestrial biota increase diversity and, if carefully located, alter distribution and velocity of flow patterns in bottomland hardwoods.

Site topography also influences runoff patterns, especially amounts and patterns of flow. Both spatial distribution and timing of runoff vary with drainage patterns because of elevation differences between different portions of the watershed. If an adequate water supply is uncertain for the size and type of wetlands selected, individual ponds or cells may be sized and positioned to receive runoff from different portions of the contributing watershed. Topography within the site and in adjacent areas must also be considered in terms of access for construction equipment and future operation and management needs.

Topography may enhance or negate the functional values expected from the new wetlands. For example, a productive, well-developed marsh may receive little use by waterfowl if it is a linear system nestled in a long, narrow ravine or valley. Many waterfowl are reluctant to rest or feed in areas with views that restrict their ability to detect approaching predators. Some diving ducks, grebes, and loons may have inadequate take-off distance and the wetlands could become a trap, as happens when they mistake a rain-slick roadway or parking lot for deeper water. Similarly, a linear wetlands perpendicular to the main stem of a river will be less successful for flood water storage since high flows will tend to bypass the wetlands rather than topping the river bank and spreading into a system adjacent and parallel to the main channel. A round or elliptical wetland may cause short-circuiting and reduce the effective treatment area in a wastewater treatment wetlands or increase evaporative losses in a groundwater recharge system.

Soils

Soil type and erodibility and composition of underlying materials will also influence wetlands design. Foremost is the permeability of the substrate, since trying to impound water over pervious soils may require excessive compaction or lining. Conversely, locating a groundwater recharge wetlands on impermeable clays is not likely to achieve project objectives. Placing dikes in areas with impermeable clays will reduce costs for transporting materials and reduce or eliminate dike seepage. In some low areas, soil materials carried in over time have created an impermeable bottom layer that is ideal for most wetlands. However, the underlying material might be highly pervious and virtually impossible to seal after the upper layer is broken through. This is often the case in wide valleys because much of the floodplain was, at one time, part of an old channel during which sand and gravel was deposited. As the river meandered away, only fine clays were deposited during floods, creating an impervious seal above very porous materials. Very shallow excavations and importation of clay dike materials may avoid penetrating the bottom seal, but there is always a potential for major leaks stemming from future activities.

Highly erodible soils in the watershed above the new wetlands may require a deep retention/sedimentation pond above the wetlands to reduce deposition within the system. On the other end, water control structures and spillways will need careful design to reduce potential downstream erosion. A well-developed topsoil on-site should be removed, stockpiled, and later spread in the graded cells to accelerate development of wetlands vegetation and substrates. Poor clay or sandy soils should also be saved since they are probably better planting media than parent material, but they may require adding lime, fertilizer, or organic materials to improve growing conditions. Very poor soils may also affect plant survival and propagation, thus impacting the choices of plant species and prolonging the start-up period.

Unless a long-term cooperative agreement has been developed with adjacent landowners, location of property lines will constrain size and location of the wetlands. Unfortunately, property lines rarely follow natural landscape features, but instead are based on arbitrary conventions (the township and range system in much of the U.S.) or historical use patterns (metes and bounds of the eastern states). Dam locations and future water levels must be related to on-site and off-site land elevations so that adjacent lands will not be flooded during normal operation, periods of high runoff, or when beavers assume management of water levels. Impounded waters should not be permitted to cause land subsidence, shoreline erosion, or become a "public nuisance" or "deadly attraction" that may jeopardize neighboring children or livestock. Opinions of adjacent landowners should be solicited to identify fears and, if feasible, accommodate desires. Remember, creating an alligator habitat in a Florida subdivision or mallard habitat among the small-grain farms of North Dakota will not endear you to your neighbors.

Configuration of the watershed may suggest that a series of small cells located on different but proximal drainages would optimize available water supplies or runoff and roughly approximate the desired total acreage. For example, a small wetlands in the mouth of each small ravine or valley in the upper portion of a watershed may, in the aggregate, create more total wetlands acreage at less expense and with lower water requirements than one large system in the lower reaches. The former system may be easier to plan and construct, depending on the relief of the upper watershed; and it will be less susceptible to sedimentation during stormwater runoff. However, if runoff is the sole water source in arid regions, the combined runoff from much of a large drainage may be needed to sustain water levels during the dry period and the lower site would be favored.

SPECIFIC DESIGNS

Access: Overlooks and Boardwalks

Lakes and impounded waters, even without the added features of complex, diverse biological communities in wetlands, tend to attract human visitors. This is especially true with society's current interest in wetlands. Consequently, visitor accommodations should be included in the wetlands design unless system function or integrity precludes other uses. Encouraging ancillary uses will often improve public support for creating wetlands for many other purposes. For example, at a recent workshop on constructed wetlands for wastewater treatment in Cannon Beach, OR, one town citizen explained that over two thirds of the community attended the ribbon cutting ceremony for the town's new wastewater treatment wetlands. When I asked treatment plant operators in the audience, "how many residents of their towns knew the location of their wastewater treatment plant?", most said almost none and one did not think the mayor or council could find his. In Cannon Beach, many residents, tourists, and a few elk enjoy the quiet footpaths through a large wetlands a few blocks from Main Street; and they support the operation and budget needs during critical periods.

Public information and education programs should be considered in planning, design, and operating stages. Certainly, public information campaigns and outreach programs will be vital to acceptance and long-term success of this technology and should be a major element of any planning effort. Encouraging public visits to view the processes at work will enhance a project's acceptance, win over valuable allies, and garner broader public support for using constructed wetlands for wastewater treatment.

Part of this effort should develop support from local sportsmen's organizations; environmental groups; naturalist, birdwatching, and wildflower societies; and other associations and individuals interested in wetlands and wildlife. Constructed wetlands may prove to be excellent areas for hunting, birdwatching, and other outdoor recreation. For example, a 1988 issue of *Birder's World,* a national birdwatching tabloid, has a 5-page article on California's Arcata Marsh and Wildlife Sanctuary. Entitled "Birding Hot Spots," the story describes the history of the town's wastewater treatment project and over 200 species of birds and 1000s of waterfowl and shorebirds using the marsh. User surveys of visitors to the Arcata wetland rank walking and isolation as important values. Widening the support circles to include average citizens will improve acceptance of this important technology.

Visitor accommodations must provide reasonable protection from possible dangers that may be encountered or the owner may fall victim to one of our litigatious society members. However, "reasonable precautions" may be differentially interpreted and even the careless that manage to drive off a dike into the pool may be successful in some courtrooms. Generally, preventing access to known hazards, correcting hazards forthwith, or taking reasonable precautions will avoid most problems that could result in legal liabilities. If the condition could be construed as dangerous or perilous, either correct it or close the area.

Visitor facilities range from occasionally mowed footpaths to elaborate visitor centers with creative displays, multimedia theaters, and full-time interpretive staff. The latter may be found at a few National Parks (Everglades) and Wildlife Refuges (Okeefenokee) or nature centers owned and operated by conservation organizations (Audubon, Nature Conservancy), utilities (Florida Power and Light), private entities (Cypress Gardens near Charleston, SC), and local and state governments (Myrtle Beach, SC). Footpaths with or without interpretive signs, and roadways with turnouts or parking at viewing points will provide basic access for most general-purpose ancillary uses. Both should be carefully located on higher ground or dikes to avoid impacting wetlands hydrology or biological communities.

Most wetlands are flat with few hills or mounds that could furnish an overlook or vantage point; but appreciation for wetlands is hindered if the viewpoint is limited to human eye level. In created systems, dikes and water control structures may elevate newcomers above the tops of marsh vegetation, but rarely to treetop levels. Getting down close to wetland plants and animals and rising well above the system aid in understanding the fascinating complexity on the one hand and grasping the incredible diversity on the other. Carefully planned boardwalks furnish the up-close perspective, while mounds or short towers are useful in visualizing the overall system.

Boardwalks are justly popular, providing access within the wetlands while protecting the visitor and the wetlands from intimate contact. Location and construction of boardwalks must be prudently undertaken lest either harm the structure and/or function of the wetlands system. Boardwalks are typically framed across short pilings driven into the substrate and made of treated lumber. Occasionally, cypress or other naturally resistant wood is used, but creosote or chemically pressure-treated lumber is often readily available. Precautions are necessary in handling treated materials and in a few instances, CCA (copper, chromium, arsenic)-treated lumber has caused mortality to wildlife in wet situations. Since the purpose is to expose the visitor to wetlands diversity, boardwalks should transect as many different types of habitats as feasible, but should not follow boundaries between zones lest they become barriers to animal movements. Linear or angular patterns are easier to construct but much less likely to blend with the surroundings than shallow curves, loops, or circles. Attention can be focused on points of interest or vistas by widening the walk, adding a bench, or putting in a short tower.

In marshes or bogs, overlooks need only be a few feet above the surface to furnish an educational vantage point. Building up a small portion of a dike or forming a mound or low hill with excess spoil material along a border may be slightly more expensive initially, but will require little maintenance and cost over the years. In addition, mounds are less likely to become safety hazards through accident or neglect and, consequently, have lower potential liabilities.

In forested wetlands, a tower may be necessary to achieve a similar perspective since the viewpoint must be at or above treetop level. Though costly to construct and maintain, the benefit to visitor understanding and appreciation of large wetlands is commensurate with the expense. Unused fire towers are popular with visitors on many National Wildlife Refuges because of the improved perspective of large wetlands from the elevated viewpoint. Unfortunately, few staff or visitors have an opportunity to observe wetlands from light aircraft at low altitudes, and providing a tower is the next best alternative.

Signs are important to visitors and to project planners and managers. Signs should not only be preclusive, but they should be informative. If closure or caution is needed, explain the requirement in terms of wetlands benefits, functions, and values, or perhaps visitor safety. Signs, small displays, and kiosks also provide opportunities for education on wetlands ecology and explanation of project objectives.

Regardless of the level of visitor accommodations, it is imperative that facilities are included for access control since visitation during certain periods may impact the principle functions or endanger the visitor. Depending on available surveillance, signed and locked gates may be adequate or sturdy steel posts and bars may be required. In any case, signs should explain the reasons for and the duration of anticipated closure and specific areas that are closed or open.

Sizing and Configuration

Natural wetlands and created wetlands vary from small backyard systems to thousands or millions of hectares. Size and shape of created wetlands are determined by site characteristics, desired functional values and funding available for land acquisition, construction, and long-term maintenance. An *Iris/Sagittaria* marsh of only 20 to 200 m^2 may purify discharges from the household septic tank and add a focal point of floral diversity to a homeowner's backyard, but only small forms of wildlife are likely to find adequate habitat. Conversely, providing significant groundwater recharge or flood water storage capacity may require hundreds or thousands of hectares as would a wetlands supporting populations of larger wildlife species.

As with so many other design factors, size and configuration are largely dependent upon the target functions, though in practice, designers tend to make the size fit the budget available for land costs and construction. Unfortunately, few consider long-term maintenance requirements or establish a funding base to cover these costs. In addition, little consideration is given to insuring that the planned size is adequate to support all of the components that enhance self-regulatory processes and reduce maintenance requirements. Current attempts to reduce size and complexity of wetlands constructed for wastewater treatment to the barest essentials are quite likely to lose an important attribute — self-maintenance — that is an important component in comparative cost analyses.

Wetlands designed for wildlife habitat must provide food and shelter — the essential ingredients for life — in adequate amounts and locations in a large enough area to provide living space for the target species. Small animals tend to have small home ranges requiring less area to support a viable population. For example, a few frogs, salamanders, toads, small fish, and turtles, and small plants (submergents such as *Potamogeton, Ceratophyllum, Myriophyllum,* floating types such as *Lemna, Azolla,* and small species of *Brasenia, Peltandra,* and *Sagittaria*) could thrive in a backyard or school yard system and even songbirds would visit occasionally. Along with myriads of macro- and micro-invertebrates and plants, this small but complex system could function as a laboratory for formal classes, for informal visits by neighborhood children, or simply as a tranquil spot of nature in over-developed suburbia. At the smallest extreme, it would not be much different from the goldfish/lily ponds fashionable around the turn of the century. However, it would likely require quite a bit of attention to maintain the original mix and, without a zone of emergent/wet meadow plants or shrubs, even small mammals (mice or voles) are unlikely to find homes.

Attracting or providing homes for larger animals will require considerably more area and seasonal aspects become important considerations (see Figure 2). Depending on the season, habitat requirements for a single species may vary substantially. During spring and early summer, many waterfowl seek the isolation of small ponds; mallards or wood ducks may be found in marshes or ponds, some temporary, of a hectare or less. During migration or on the wintering grounds, ducks and geese congregate in large flocks that tend to avoid small wetlands, probably because of limited protection from predators but perhaps also because the large flocks simply need larger physical space. However, in many instances, waterfowl loaf almost shoulder to shoulder in a small portion of a large bay or marsh. Though they actively use only part of the total area available, they rarely frequent wetlands that are only as large as the area used in the larger system. In some cases, the small portion used changes and, over time, much of the larger system may receive some use. However, in others, only a small percentage of the total marsh or swamp is visited by the flocks. Undoubtedly, security from predators is a factor in this behavior, but other elements may be important.

Feeding behavior and food habits vary considerably, though only a few wetlands species are as restricted as some terrestrial species. Wading birds tend to eat anything that walks, crawls, or swims as long as it is small enough to swallow. On the other hand, snail kites (*Rostrhamus sociabilis*) live almost entirely on one type of snail (*Pomacea*). In general, animals with highly specialized food habits will require a much larger tract to produce adequate food supplies for a self-sustaining population than will species with generalized food habits (see Figure 3). The latter may be able to forage on a wide variety of food items including plants, insects, herps, mammals, and birds. Snapping turtles (*Chelydra serpentina*) (large and generalized) occur in almost any wet environment east of the Rockies, from small ponds and even wastewater lagoons to extensive marshes and swamps; bog turtles (*Clemmys muhlenbergi*) (small and specialized) are restricted to a few bogs with narrowly defined water chemistry and plant species.

Project goals may go beyond simply the presence or absence to setting production targets; that is, the goal may be to raise or produce a certain number of one or more types of plants or animals during a given time period. In that case, planners must review scientific literature on the desired species and determine typical production rates per

Figure 2.   A viable raccoon population needs a sizeable wetlands to provide adequate supplies of food and cover for all its members.

Figure 3.   Larger wetlands, simply because of their size, have the capability to support many more of the important functional values than do small systems.

unit area of wetlands for that species or group of species and size the new wetlands accordingly. Depending on the species, time considerations and scheduling may also become important. A harvestable crop of crayfish can be raised in 1 or 2 years on relatively small areas, but commercially valuable stands of cypress or cherrybark oak require large areas and many growing seasons.

Sizing wetlands for flood storage or hydrologic buffering is much less exact since our understanding of this function is limited. However, storage capacity may simply be determined from the water volumes, peak runoff periods, and planned retention times for average storm flows of any given period. For example, flood storage wetlands upstream from an urban area may be designed to store and slowly release excess flows caused by a 10-year 24-h storm or a 100-year 24-h storm. Total volumes and peak discharge rates may be estimated from anticipated rainfall and watershed characteristics to determine the requisite storage capacity. Wetlands size is then predicted from the water depths attainable and the area needed to store flood waters for a given time period. For example, if the predicted storm runoff volume is 1,000,000 m^3, then the wetlands should be capable of holding 1 m deep water over 1,000,000 m^2 (100,000 ha), 0.5 m over 200,000 ha, or some other combination. Obviously, this wetlands should be located in the restored floodplain along either or both sides of the river. Entry into and exit from the wetlands may be simply by over-topping the river bank, or specific channels, levees, and control structures may be needed.

The basic concept is to simply provide the capacity to store a given volume of water, to reduce the flow velocity by forcing it through dense vegetation, and to gradually release flood waters into the river or stream for an extended period after the storm. Depending on time of year, the storage and release period could range from 2 to 4 weeks or be as short as 3 to 5 days without significant damage to the wetlands. During winter, bottomland hardwoods often withstand inundation for weeks without apparent harm; but flooding for more than 4 to 5 days in spring and early summer will cause stress and mortality. A cattail or reed marsh could tolerate an additional 0.2 to 0.5 m of water for 2 to 3 weeks in the growing season, but longer flooding will cause changes in submergent and wet meadow species and eventually in reed or cattail. Consequently, storage capacity and wetlands sizing must be related to the time of year when flood protection is needed and the species used in the flood storage wetlands.

The obverse of flooding, but a part of total hydrologic buffering, is the augmentation of low river flows during dry periods; this poorly understood phenomenom is very likely a simple function of storage and release over an extended period. Doubtless, waters draining into a river from a floodplain wetlands during intervals between storms tend to maintain consistent river levels, moderating the peaks and troughs that would otherwise occur. However, wetlands drainage may vary substantially with hydraulic conductivity of the duff/litter layer, the soil layer, and the underlaying parent materials. Some flow will occur across the surface, but once the water level in the wetlands has dropped below the berm or overbank, return flow will be through channels or underground. In many cases, drainage back to the river may be largely subterranean due to the highly permeable nature of underlying materials in many floodplains.

Since the flood storage wetlands are planned as a "gentle" sort of dam, they should occupy the entire width of the floodplain in the selected portion of the valley. If this is not possible, then levees or dikes must prevent flood waters from escaping around an end or through a non-wetland zone within the floodplain. They could also be located on one side of the river if the opposite bank is naturally or artificially raised at or slightly above the planned water level elevation within the wetlands. Since current velocity is highest on the outer radius of a curve, overtopping normally occurs first on a bend and, if only one side of the river is available, an astute planner will select an area for the wetlands immediately downstream of a sharp bend with a low bank.

If flood water ingress and egress is broadly across the river bank, shape may not be important. However, if ingress and egress are through channels or old oxbows, then an elliptical, rounded diamond or V-shape pointing downstream will be most efficient in accomplishing rapid accumulation and gradual releases. In these instances, low dikes or levees may be needed to guide flood waters into and contain them in the wetlands, and small control structures will be useful in managing releases.

For wastewater treatment, wetlands size and configuration are determined by the volume of flow and the influent and effluent concentration of pollutants. Required effective treatment area is related to mass quantities of organic materials, metallic ions, or concentration of microorganisms. Conversely, retention times or hydraulic loading may be used to estimate required volumes and, indirectly, treatment areas. Though proponents of mass estimates or hydraulic loading lay claims for the merits of one vs. the other method, current recommendations using either method tend to estimate similar wetlands areas for specific influent-effluent concentrations. That is not surprising since both methods are applied conservatively at present because of our limited understanding of processes and mechanisms and because either is to some extent a manifestation of the same concept. However, the mass-treatment area method attempts to relate the quantity of contaminants to the surface area for attachment of microbial populations that decompose or alter those substances. The retention-hydraulic loading method assumes that microbial populations need a certain period of time to modify pollutants and estimates the storage volume required to constitute that minimum turnover period. The latter was developed in conventional sewage treatment because, in any one region, municipal wastewater flows have similar pollutant concentrations.

However, comparisons between regions, for example, between the U.S. and Europe, become confusing since water usage, and subsequently, pollutant concentrations vary substantially. Average U.S. water use is approximately 400 L/person/day whereas the average in England is 200 to 250 L/person/day; and in central Europe, the average may be as low as 100 to 120 L/person/day. Since the total amount of organic waste (mass) generated by one person is similar in each instance, the concentration of organic waste (proportional amount in a given quantity of water) is quite different between the U.S. and central Europe. Comparing hydraulic loading rates (retention times) is unrealistic, but comparing organic mass loading provides a common basis for evaluating loading rates and system performance. Similar loading comparison problems arise because of widely varying concentrations in industrial, agricultural, or acid drainage wastewaters.

Since most wastewater treatment wetlands have been designed for minimum size and cost to provide the required level of pollutant removal, maximizing effective treatment area and reducing short-circuiting or unused treatment area results in rectangular shapes. Generally, inlet distribution and outlet collection piping is located completely across the upper or lower end of the cell and a rectangular shape theoretically enhances broad sheet flow across the width of the cell. Circular or elliptical shapes would likely have unused portions in each cell, but a V shape pointing downstream might provide an alternate design without creating short-circuiting. It may not be too effective for nitrogen removal, but should be tested for biochemical oxygen demand (BOD) and suspended solids (TSS) removal. A few projects have used linear systems contoured into the hillside because little level land was available and these tend to blend into the surroundings better than the standard rectangle. Generally, wastewater treatment wetlands should have a 3-4:1 length-to-width ratio and rectangular shape if minimal treatment area is available. However, other more aesthetic configurations would be practical if the wetlands area was increased and ancillary benefits are also likely to increase with larger size.

Larger size has an added benefit in that contaminant loading will be less, loading fluctuations more readily accommodated, and greater diversity of plant and animal species possible, hence improving the water purification function as well as increasing other beneficial functions. Although planners tend to design the smallest wetlands with the fewest components, that is, reduce the treatment system to the bare essentials, for most wastewater treatment projects, treatment efficiency and system resiliency increase with added size and biological complexity. Small simple systems are vulnerable to upset from fluctuating loading rates or pest outbreaks in important vegetation components. Too small and too simple will likely cause higher operating and maintenance costs and hamper system performance.

Lastly, the new wetland should be designed to blend in with other features of the landscape, including those on adjacent lands under other ownership. Within the constraints outlined above, shape and dimensions of the new system will be determined by the principle functions. However, dams across small drainages tend to form teardrop shapes and excavated ponds/wetlands tend to be rectangular because of earth moving equipment procedures. Neither adds much and may even detract from the aesthetics of the landscape; but adding a small water body (a pond/wetland) can significantly improve landscape aesthetics with a little foresight and planning. Regular shapes (angular, rectangular, or round) should be avoided if at all possible. Smooth, rounded, elliptical, or sinusoidal curves and irregular lines form pleasing shapes that mimic and merge with natural features of the landscape, thus augmenting visual attractiveness and habitat diversity.

If regular configurations must be used, appearance of a rectangular excavation may be improved by selective grading of shallow areas along the boundaries to create irregular shorelines and increase edge effect. Spoil islands, unequal and uneven location of deep and shallow water regions, patchwork patterns of different plants and different trees, plants with different heights and growth form, slow and fast growing trees, careful location of observation points and vistas, plants or trees with colorful fruits or foliage, showy flowering species, smoothly contoured dikes, and screened control structures can all reduce the obtrusive, "engineered" appearance of a new wetlands.

Sealing and Lining

Most natural wetlands occur in topographical lows because water runs downhill and accumulates in depressions. However, all depressions do not support wetlands ecosystems. Many basins are directly connected to subterranean channels (karst sinkholes) and others lack an impervious substrate, allowing potential water supplies to infiltrate to ground water. Without ponding or flooding during the growing season, these sites support, at best, transitional and, more commonly, terrestrial ecosystems rather than wetlands. Even though adequate water, energy, and nutrients are carried off adjacent uplands, the uninterrupted connection to ground water permits passage of these critical elements through the system and into ground waters. Conversely, depressions with wetlands have a barrier to reduce or prevent water losses to the underlying materials and, consequently, undergo extended periods of inundation required to exclude terrestrial plants and animals. Only a few exceptions exist wherein inflows (surface and/or subsurface) balance percolation losses and other losses and the basin nurtures a wetlands ecosystem without having an impermeable lining. Most natural wetlands are perched above an impervious layer that reduces or prevents water loss to underlying strata, as occurs in upland environments.

A few planners may be fortunate in having a continuous, reliable source of clean water from a spring or permanent stream to offset subsurface losses, and insuring the new wetland has an impervious layer may be less critical. Others may compute a balance of inflows from seepage or perhaps wastewater flows and losses from all factors and conclude that moderate or large infiltration losses are tolerable. In the first case, through flow could result in the loss of important nutrients and energy and the new wetlands may be less productive than one that continuously cycles nutrients. In addition, even with relatively clean sources, introduction of unmodified compounds originating in uplands or within the wetlands to groundwater may detrimentally impact groundwater quality. Conversely, if the reliable water supply is any type of wastewater, isolation from groundwater is imperative in order to protect groundwater, even if adequate water is available to sustain a leaky wetlands complex. Though water quality near the discharge end of the complex may be high and not perilous, insuring that leakage only occurs at the lower end may not be practical, and partial protection is justifiably discouraged by regulatory agencies.

Since an impervious liner is so important, designers must evaluate composition of soils on site and determine the hydraulic conductivity with percolation tests. Soils in many otherwise suitable sites may lack adequate clay content or the particle size is too large to produce a natural seal. If the conductivity is greater than 10^{-6} or 10^{-7} cm/s, sealing or lining must be considered, especially in arid climates, wastewater treatment systems, or other situations where loss from infiltration could jeopardize water level maintenance during extended drought periods.

Compaction of *in situ* soil materials is the simplest and least costly method of sealing the bottom, but the clay content must be greater than 10%, and a wide range of silt, sands, and other small particle sizes should make up the majority of the soil. During construction, the bottom is tilled to a 30 to 40-cm depth and then rolled with a sheepsfoot roller under optimum moisture conditions to create a dense tight layer of the same depth. If insufficient clay is present, more suitable material (clay content >20%) may be imported from a borrow area and layered across the bottom and up the sides of the dikes to operating water level elevations and compacted as above.

If permeability is high or borrowed clay unavailable, soda ash or bentonite clay may be obtained through the nearest farmers supply or direct from the source. Bentonite is typically applied at 1 to 3 lbs/ft^2 and soda ash at 0.10 to 0.20 lb/ft^2, depending upon the results of laboratory analysis of native soil materials. Bentonite is used if in situ materials have inadequate clay and are mostly coarse-grained particles. Very fine-grained clay soils may also permit substantial seepage, in which case soda ash (a chemical agent) is used as a dispersing agent to rearrange clay particles and seal the bottom. Both are layered uniformly across the bottom and up the dikes, tilled in to 40 to 60-cm depths, and compacted with a sheepsfoot roller before flooding. Bentonite is a colloidal clay that swells many times its original volume when wet, but shrinks again upon drying. Consequently, it must be kept wet after application and should not be used if significant water level fluctuation is expected or complete drying will occur. The compacted soda ash layer must be protected against penetration by roots, animals, or from erosion by covering it with a 40 to 60-cm layer of soil or rock near inlets and spillways. Bentonite typically retails for $200.00 per ton and soda ash for $350.00 per ton at local outlets, but substantial discounts on large quantities can usually be obtained directly from the source. However, if large amounts are needed and/or the source is distant, materials cost plus freight charges may become prohibitively expensive.

If bentonite sources are not nearby and/or the required application rate becomes greater than 1.5 to 2 lbs/ft^2, comparative costs with synthetic materials become less attractive. Geotextiles are extensively used for a variety of sealing purposes and generally available in most areas, though relatively expensive to purchase and costly to install. Depending on thickness and composition, most synthetics are susceptible to damage from root growth and must be carefully placed below a 40 to 60-cm layer of soil materials. Proper installation also requires placing a layer of fine sand to cushion the synthetic, and applying special adhesives to weld long strips together to form impermeable seals. Most synthetics are also vulnerable to UV radiation damage and must be covered with at least 10 to 15 cm of soil where ever used. Since installation is critical to obtaining a good seal, it should not be undertaken by the inexperienced.

Irrespective of the method used to create an impermeable liner, 40 to 60 cm of soil should be placed above the liner to support planted vegetation. Roots of few if any wetland plants will penetrate deeper than 40 cm unless severe prolonged drought conditions occur. Providing adequate substrate for root growth decreases the potential for root penetration of the liner and subsequent leakage.

Most of these methods could be used in constructing smaller wetlands, but attempting to seal hundreds or thousands of hectares in a large system is usually impractical unless a bentonite mine is adjacent to the site. If the soils on the site of a large system are pervious, it is generally better to find another site since even compaction of in situ materials over a large area could quickly become the most expensive element in project costs. However, regardless of the means, the created wetlands should have a relatively impervious liner to preserve water supplies, foster nutrient cycling, and protect groundwaters.

Dikes, Dams, and Berms

Restoring wetlands may require little dike construction, but creating a new system will likely require building one or more dams or berms if elevations are relatively flat. In any case, these structures function to modify site hydrology by restricting water flow to create a pool of standing or slow moving water that supports the biotic communities.

In general, dikes will be fairly low (see Figure 4). If heights are greater than 2 m or failure could cause serious damage downstream, planners should employ full engineering design procedures or should secure the assistance of an experienced engineer to insure dam integrity and safety under anticipated conditions. With lower dikes or berms, caution is necessary, but design and construction are less complicated and feasible for less experienced planners as long as catastrophic failure would not cause legal liabilities.

The foundation for the dike should consist of impermeable clay material or bedrock without crevices, seams, or fissures. If the location has sand or gravel, it may be useable, but the assistance of the local SCS engineer or another engineer experienced with local conditions should be obtained. Bedrock should be carefully inspected to insure that excessive seepage will not occur through seams or cracks and that bonding can be obtained with dike materials.

If available dike material is not impermeable clay, a clay core should be installed in the dike, insuring that the core is well bonded to an adequate foundation. Slope on the sides should be a minimum of 3:1 with top width adequate for vehicular travel. On a very small system with low dikes, widths may be 1 to 1.5 m to at least provide space for mowing and foot paths. Larger systems need top widths of at least 3 m to accommodate vehicles of operating personnel and mowing equipment. Dikes used by visitors should not be less than 4 m, and preferably 5 m wide. However, as top width increases, width of the base increases proportionately and a considerable amount of wetlands area may be lost to dike networks. For example, a 2-m high dike with top width of 3 m and 3:1 slopes will have a 15-m base.

Muskrats often burrow into dikes and can cause dike failure unless preventive measures are included in the designs or problems are identified and corrected quickly. The latter, in some areas, could become an expensive operating procedure. Wide dikes (3 to 5-m top widths) with 4:1 or greater slopes and corresponding bases rarely have serious muskrat damage. Muskrat tunnels and dens generally do not extend into the dike over 1 to 1.5 m. However, burrowing, in some cases from both sides, into narrow or steep dikes often causes collapse of the surface and may cause total failure. In small systems, installing welded wire vertically in the dike during construction will prevent muskrats from burrowing through the dike and causing failure. Rock rip-rap placed on both surfaces from pool bottom to 0.7 m above normal water elevations will also discourage muskrat burrowing, but it inhibits vegetation and impacts aesthetics.

Water Control Structures

Dikes or dams retain and impound water to form the hydrologic environment for the new wetlands, but some means of manipulating the water elevations behind each dike

Figure 4.   Dikes in Bear River marsh not only regulate water levels in upstream pools, but also protect freshwater marshes from salt water influences when water levels rise in Great Salt Lake.

is essential to wetlands system management. Most created wetlands will require deliberate management—generally, water level manipulation—especially during the first 2 to 3 years to encourage vegetation establishment (see Chapter 12). Once the plant and animal communities are in place, little management will be required until undesirable successional changes begin some years later. Typically, these are represented by drastically declining productivity or by invasions of terrestrial species. Constant water levels for many years in most marshes will cause production decreases because essential nutrients are immobilized in reduced states within the substrate and not available to the plants. Complete drying for an extended period during the growing season (drawdown) oxidizes and mobilizes substrate nutrients so that an explosion of growth often follows reflooding.

Terrestrial plants can only invade an older system because water depths are gradually decreasing from accumulation of humic materials, peat, and sediments. Moderately increasing water elevations during the growing season will inhibit or eliminate terrestrial invaders without severely impacting wetlands plants. However, over time, the required level may exceed the elevation of the top of the water control structure or dangerously reduce remaining freeboard on the dikes. At this point, complete drying during the growing season and perhaps burning the accumulated peat under appropriate conditions and with necessary permits will be needed. In severe cases, dredging of the accumulated materials may be required.

Some wetlands have been designed with only an overflow spillway to prevent excessive water levels that may damage dikes. Many constructed wetlands treating acid drainage in remote areas rely solely on overflow spillways because water control structures are expensive and susceptible to vandalism. A few small constructed wetlands treating livestock waste or row-crop runoff also lack control structures for similar reasons. In these cases, normal biological productivity and complexity are subordinated to the primary purpose of wastewater treatment and even invasion of terrestrial plants as the depression gradually fills is acceptable. However, management of these systems is only possible with earth-moving equipment and little can be done to provide the critical disturbance element that would retard the inevitable succession to a terrestrial system. At that point, it may be necessary to breach dikes, dewater the system, and dredge out the accumulated deposits followed by dike repair to restart the wetlands.

Overflow spillways in small, low-flow systems often consist of a low portion in the dike with dense coverage of water-tolerant, mat-forming grasses, sedges, rushes, or rock riprap. During construction, a porous geotextile fabric or one of the various organic mattes is placed over the spillway to reduce erosion until planted vegetation has become established. Repair of minor erosion may be needed after unusual storms, but little investment is jeopardized and repair costs are insignificant. More rugged protection may be obtained with riprap or a layer of concrete placed within and along the sides of the spillway. The extreme is a fixed elevation, concrete structure that is much more costly to build and provides little additional protection or control though a few have been used in larger systems.

Depth and duration of flooding control much of the plant community and, directly and indirectly, the animal community in wetlands ecosystems. Consequently, wetlands management primarily consists of water level manipulation, and water control structures in the dikes are essential tools for managing created wetlands systems. Though a variety of designs are available, keep in mind that the principle use will be for setting a specific elevation to maintain a desired water depth in the upstream pool. In some types of established wetlands, that elevation may not change during the year or over the course of many years. In others, it may be necessary to raise or lower the level during the growing or non-growing seasons to simulate natural hydrologic cycles. Infrequently, the water control will be used to drastically lower and perhaps gradually raise the pool level to foster germination and establishment of wetland plants. It may also be used to completely drain the pool for dike repair or other needed maintenance, or for deep flooding to retard or reverse successional changes. Generally, a certain water depth will be maintained for weeks, months, or even years on end and the choice of water control structure design should reflect expected operation and the principal function. The ideal water control structure will:

1.   provide for fairly precise regulation of water elevations
2.   have the capacity to raise water levels to the maximum permissible level with a safe margin of dike freeboard, essentially the elevation of the emergency spillway outlet
3.   have the capacity to completely dewater the pool
4.   allow changes to be easily made
5.   not require changes because of increases or decreases in inflows or from precipitation
6.   consist of simple structures requiring little or no maintenance
7.   not be susceptible to vandalism
8.   not be susceptible to blockage from debris or plant growth
9.   inhibit blockage by beaver or muskrat

Available designs do not meet all criteria completely, though some are more appropriate to certain circumstances than others. Various types of valves or penstocks are perhaps most commonly used and are most suitable for very large structures with high flow capacities. Valves, whether gate or ball, can be installed in virtually any diameter of pipe and flows are regulated by partially obstructing the opening (reducing the functional diameter of the pipe). Large valves often have stem screws or other mechanical devices to facilitate raising and lowering the gate despite considerable hydraulic pressure. With a given pressure and above very small volumes, valves will provide accurate regulation of flow *volumes*. At minimal flows, valves are susceptible to clogging that may require frequent opening of the valve to its maximum to flush the blockage.

Valves are designed to regulate *volume* of flow, not water level *elevations*. With experience, it is possible to determine the appropriate setting to balance outflow from a wetland pool with normal inflow and even to learn what degree of adjustment is necessary to compensate for a certain amount of rainfall. However, many mistakes will be made in developing this experience and the water control must be adjusted after every moderate or heavy rain and then again when the increased inflow has passed through the system. Conversely, it will also need to be adjusted, perhaps repeatedly, during a prolonged dry period. Though valves are suitable in high flow systems, they require considerably more adjusting than simpler methods because valves regulate *volume* of flow, not water *levels*. However, if water depths are much more than 1 m in the lower portion of the pool, other methods may be impractical or even dangerous to adjust and valves should be used.

Two types of water controls designed to regulate water elevations have been in use in smaller systems for many years. Perhaps the oldest and most widely used is the "flashboard" or "stoplog" type depicted in Figure 5. Concrete reinforcing rod pins on each side of the log are made to engage a simple fork of similar material to raise or lower each log.

The control structure is normally built of reinforced concrete within the dike at the lowest end of the pool and the floor of the opening is located at or below the bottom the wetland pool. Height of the opening (<2 m) is scaled to the maximum depth desired within safe limits of dike freeboard and, commonly, the top elevation is the same as the elevation of the emergency spillway. Width of the opening is determined by the cross-sectional area required to completely drain the wetland pool in a specified period — usually 3 to 5 days. The opening width should also accommodate increased flows projected from runoff following a 1-year 24-h storm event. Accommodation of greater flows is incorporated in the width and configuration of the emergency spillway. If a width of more than 1.5 to 1.7 m is needed, 2 or more banks of slots to hold logs with concrete islands (dividers) may be used. Greater widths in each opening are not practical since 4×4 or 6×6 logs will not retain their shape against water pressure over greater lengths. Alternatively, more than one individual structure may be placed in the dike if a larger total opening width is required.

Concrete wingwalls on upper and lower sides of the dike protect the dike from erosion and reduce seepage around the control structure. If the substrate of the pool is not impermeable, a vertical concrete wall may be poured below and perpendicular to the floor of the opening and parallel to the dike to prevent seepage and/or erosion below the control structure. The anti-seep wall is in turn embedded in the clay coring of the dike.

Water level regulation is obtained by placing "logs" or "boards" in the control slots to the desired elevation. One flashboard or log designed to fit in the structure slots is shown in the foreground of Figure 5. Normal inflows raise the pool level above that point and the excess simply flows over the top board. Excessive inflows from heavy rain elevate pool levels, and larger volumes flow out since the depth of flow across the log is commensurately greater. Conversely, reduced inflows to the system cause the pool level to fall below the top board, suspending outflow until pool levels are again raised by rainfall or increased inflows. Consequently, adjustment and/or readjustment is unnecessary and operation is minimal.

Major water level adjustment is simple on the one hand and could be more difficult on the other. Large increases in depth are obtained by simply placing more logs in the slots to reach the desired level. However, major reductions may require stepwise removal of boards since additional boards beyond the first 2 to 3 will have substantial horizontal and vertical water-pressure holding them in the slots, in addition to the weight of the water soaked boards. If all of the boards must be removed concurrently, a block and tackle or tractor-mounted front end loader may be the only safe, practical method of extracting the boards. Otherwise, 2 to 3 boards may be removed and, after the pool level has fallen and pressures are reduced, additional boards are removed and the process repeated until the last board is out.

Construction of a stoplog control requires considerable labor in building forms, tieing reinforcing rod, and pouring concrete, and costs are higher than for a control that

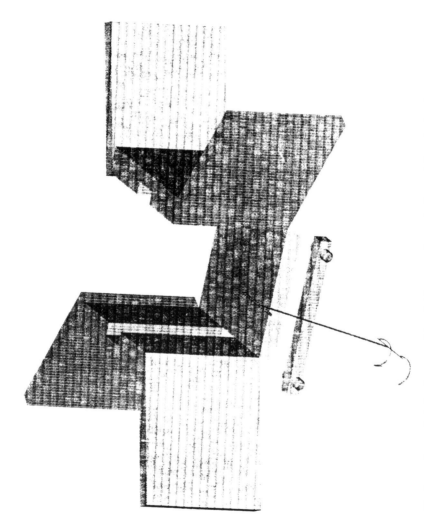

Figure 5.   Stoplog water control structures provide reliable means to precisely regulate water levels.

is fabricated offsite and set into the dike during construction. In addition, transportation of lumber and concrete materials may be expensive or difficult in remote or inaccessible areas.

The stoplog structure is more expensive than the modified version described below or the swiveling pipe design, but it has greater capacity for large flows and is less susceptible to debris blockage than either the pipe or culvert designs. Though muskrats rarely block this type unless water depths have become very low because of accumulated deposits, beaver are adept at placing chosen materials between the logs and walls and raising subsequent water levels to attain their objectives. If beaver are allowed to continue managing the control structure, over a period of time the large amount of imported material may be so tightly interwoven that extraction without explosives is a long and arduous task. Human vandalism may be reduced by securing logs with metal straps or mounting bolts and lock fittings adjacent to the slots.

A flashboard fitting modifying the simple metal pipe riser (flashboard culvert) provides considerable flexibility in water level manipulation and is generally available at the local metal culvert fabricating plant, as shown in Figure 6. Flashboard fittings are rarely placed on culverts with more than 0.7 to 0.8 m diameters, so this structure is not suitable for large systems or systems with greatly fluctuating inflows. Asphalt and other coatings are normally available if system waters are corrosive, as in some acid drainage treatment wetlands; an anti-seep collar should be included in the purchase specifications. The horizontal pipe must be placed in a trench below the floor of the wetlands pool so that the bottom of the flashboard opening is at the level of the pool floor to insure complete drainage capability.

Operation of the flashboard culvert is similar to the stoplog control structure in that boards are simply placed in the slots to the desired height and subsequent water depth in the wetlands pool. Since the boards are shorter and thinner, extracting all the boards at one time is generally feasible. However, lifting water-logged boards from the unstable platform of a small boat, if the structure is in deep water, may require considerable strength and dexterity. Operation and maintenance is minimal, though the smaller size opening is more likely to have problems from debris blockage, in which case a trash rack of steel rod or flat iron built around the pipe may be required. Since the upright pipe with the flashboard fitting is located in the deeper water of the pond and some distance out from the dike, this type is less susceptible to vandalism, though the standard design is easily blocked by beavers.

The water control enclosure in the background of Figure 6 is an anti-beaver modification to the flashboard culvert that has been quite successful. Several of these designs have been in use in beaver-inhabited wetlands for over 10 years without obstruction. The enclosure may consist of a larger diameter culvert or, in some cases, a 55-gal drum that is placed over the top of the upright section of the flashboard culvert. Holes are cut in the sides near the bottom and 3- to 5-m lengths of 10 to 20-cm plastic pipe are securely fitted in the holes. A metal cover secured with chain or metal strapping prevents beaver access through the top of the enclosure.

Beavers apparently detect "leaks" that need plugging, primarily by the sound of water flowing and perhaps secondarily by sensing flows. The principle underlying the design of this enclosure is to prevent beaver access to the flashboard fitting. Although beavers doubtless hear the water flowing over the boards and splashing into the pipe

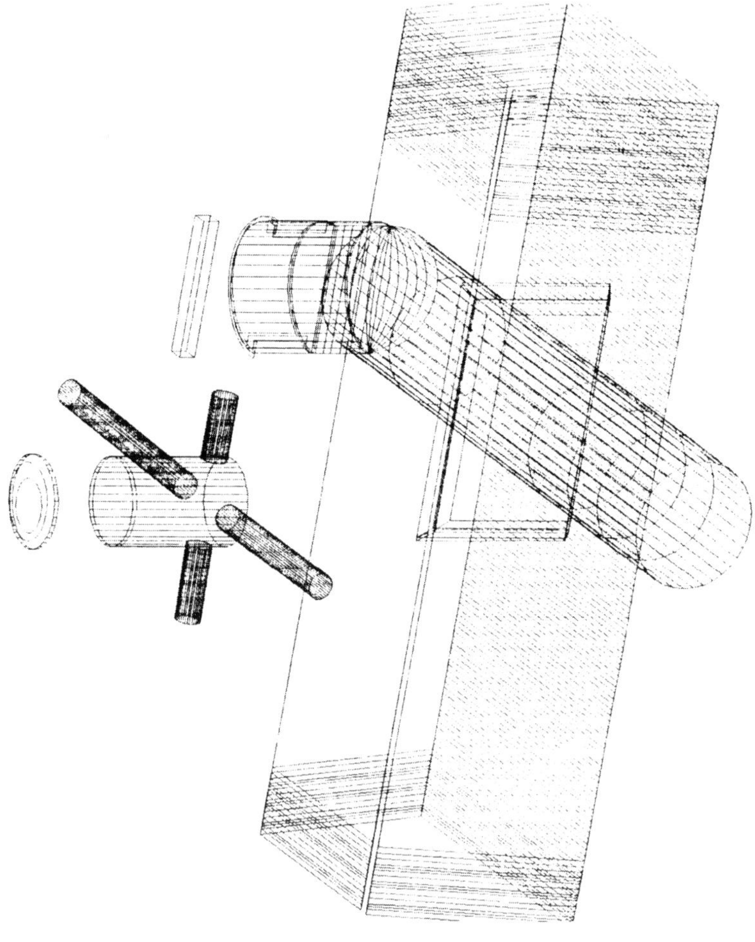

Figure 6.   Culvert-type flashboard water control structures can be made relatively beaver-proof with the enclosure shown in the background.

below, the enclosure prevents them from reaching and blocking the source of the sound — the flowing water. Since it is necessary to avoid creating rushing water sounds or to create an obvious flow pattern at the distal ends of the plastic pipe, the pipe diameter must be large enough, and 10 to 20-cm pipes should be used to prevent beavers from detecting and blocking flow into the pipes. If the pipes are too short, beavers may accidently block the ends in attempting to stop the water flow over the flashboards within the enclosure.

This anti-beaver enclosure has been successfully used for many years in created wetlands and has even been installed in beaver dams. However, given the adaptability of the beaver, some individuals will doubtless learn to manage this design as they have most others.

The swiveling pipe concept is the least expensive and perhaps the most simple structure providing excellent capacity for water level regulation (see Figure 7). Constructed of plastic pipe, it is inexpensive, readily available, and easily transported to and assembled in remote locations. The slotted or perforated collector pipe on the upstream side of the dike is often placed in a trench and covered with coarse gravel or rock, hence, the discharge pipe must also lie in a trench below the dike. Alternatively, the collector side could consist of a short riser pipe or a flush-mounted, hooded inlet pipe. Either of the latter two will require a trash rack and perhaps anti-vortex baffling on the end of the pipe. The swiveling section on the lower side is an L-shaped piece of slightly larger diameter pipe fitted over the smaller pipe with a lubricated O-ring fitting. The pipe section beneath or through the dike should have an anti-seep collar, as for metal culverts or corrugated pipes.

In operation, the elbow or L-shaped pipe is pivoted up or down by inserting a short length of wood board into the end of the pipe to increase leverage and carefully rotating the pipe. Raising or lowering the pipe outlet establishes the water level in the wetland pool behind the dike. In the vertical position, it will maintain the greatest depth, whereas horizontally it will drain the pool if the pipe joint is at or below the floor of the pond. Depending on the tightness of the fitting, friction is normally adequate to hold the L-shaped pipe at the desired angle and outlet elevation though, occasionally, it may need to be fastened with a short length of chain. A concrete pad or rock riprap should be placed around and below the discharge pipe to reduce erosion and, if flow measurement is needed, the pipe should discharge into a small well with a weir plate fitting at one end.

The swiveling pipe structure is inexpensive; pipe, fittings, and cements are readily available and easily assembled, and operation is minimal. However, this type is limited to small flow systems since pipe diameters much greater than 25 to 30 cm are impractical to rotate and it should not be used where inflow is expected to vary significantly. Generally, the discharge is equivalent to the diameter of the pipe and additional inflow with greater depths and higher hydraulic pressures only increases the velocity of outflow. Increased velocity translates into increased volumes, but the capacity to accommodate substantial increases is obviously limited. Since the inlet structure is below the pool water surface, it is not susceptible to debris blockage though accumulated humic materials and other deposits may eventually clog the slotting or perforations. Beavers can easily block a horizontal collector pipe within the pond, but may be frustrated by using an anti-beaver enclosure over a short pipe riser. This

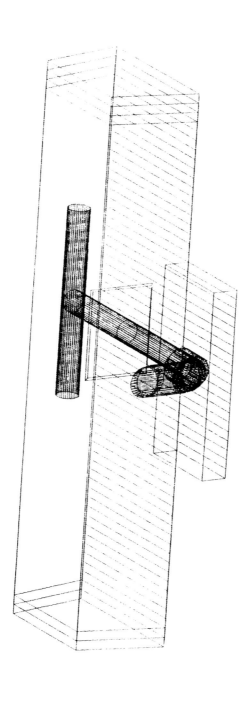

Figure 7.   Swiveling elbow-type water control structures are inexpensive and suitable for small flow systems.

swivelling pipe is the most vulnerable to vandalism since inappropriate pipe rotation is not difficult and plastic pipe can be easily damaged unless the swiveling discharge pipe is located in an enclosed concrete well.

Overall, the stoplog control structure is the most durable and trouble-free, though more expensive to build than the other types. With smaller and less fluctuating flows, either of the other designs will suffice though the flashboard culvert will accommodate more flow variability and is more amenable to protection from manipulation by beavers or damage from vandals. Valves require excessive adjustment, are expensive, and are vulnerable to beavers and vandals, although they are capable of managing higher flows than the less complex designs.

At the extreme, if the created wetlands is located in an area with low base inflows but very high inflow from intense storms, a large upstream watershed or high runoff from an unvegetated or largely impermeable watershed, a combination of designs including the stoplog and/or the flashboard culvert design might be used along with a valve structure. Unfortunately, storm events are as likely to occur at night as during work hours and valve adjustment may not be practical until after water levels in the wetlands have risen significantly. Perhaps the least complex solution for construction and operation is one or more stoplog designs coupled with careful design and maintenance of the emergency spillway. High runoff from an unvegetated watershed is also likely to carry substantial quantities of suspended materials and the designer should consider including an upstream retention/sediment pond to avoid high sediment accumulations that would have serious impacts on the wetlands. In this case, water controls in the retention pond may regulate and stabilize flows into the wetlands, thereby reducing the need for large-capacity structures in the wetlands.

If beaver problems are anticipated in a large wetlands or one with high flows, two or more flashboard culverts with anti-beaver enclosures will provide flow capacity as well as protection from beaver management of water levels. Flashboard culverts or stoplog structures are more costly and transporting materials to remote sites may be more difficult, but their longevity, limited need for adjustment, and resistance to blockage by debris or beaver make them advantageous at sites with little access control or maintenance capability.

In summary, the choice of design is likely to be strongly influenced by size of the wetlands and flow volumes since cost comparisons must incorporate long-term operation and maintenance requirements which often exceed initial construction costs. The more costly water level regulation devices have lower adjustment requirements and reduced maintenance costs because of the design simplicity and durability of the construction materials (see Table 1). Reinforced concrete and heavy corrugated metal pipe are more resistant to corrosion/decomposition and to vandalism and beaver impacts than plastic pipe or complex valve structures. In the final analysis, the type selected must have adequate capacity to regulate wetlands water elevations with minimal need for adjustment and maintenance for anticipated base flows as well as limited stormwater runoff. The stoplog design and the flashboard culvert, sized for expected volumes, have the best combination of attributes to satisfy requirements to regulate water levels as the principal means of managing the created wetlands.

Table 1.   Comparisons of Water Control Devices

Attribute	Stoplog	Flashboard Culvert	Swivel Pipe	Valve
Flow capacity	Moderate	Low	Low	High
Durability	High	Moderate	Low	Moderate
Water level regulation	High	High	High	Low
Adjustment requirements	Low	Low	Low	High
Ease of adjustment	Moderate	Moderate	High	High
Debris blockage	Low	Moderate	High	Low
Vandalism vulnerability	Low	Low	High	High
Beaver vulnerability	Moderate	Low (w/encl.)	High	Moderate
Construction cost	High	Moderate	Low	High
Complexity	Low	Low	Low	High
Material availability	Moderate	Moderate	High	Low
Materials transport	High	Moderate	Low	Low

Emergency Spillways

Since watershed runoff estimates suffer from inexact weather predictions as well as watershed surface and vegetation characteristics, including the capacity for all possible flows in the selected water control structure(s) is not only costly but impractical. The need to relieve excess water pressure for dike protection during and after high rainfall or snow melt runoff is crucial to long-term success of the wetlands. If the dikes are washed out by excess flows, water level regulation is obviously lost until possibly costly repairs are accomplished. Depending on time of year and duration of water loss, wetlands plant and animal communities may suffer serious impacts. Therefore, an emergency spillway must be included in every dike to protect dike integrity and the dependent biological communities.

The function of the emergency spillway is to provide extra discharge capacity in addition to the flow capacity of the water control structure during high runoff events. It should pass excess water over or around the dike so that ponded water levels do not overtop and damage the dike or water control structures, and excess flow is discharged without damaging either side of the dikes or control devices or the outlet channel below the water control structure. Though its operation may be infrequent and for limited periods, failure due to inadequate design or poor construction and maintenance will jeopardize the wetlands system.

Dimensions of the emergency spillway are determined from the size and nature of the contributing watershed and the storage capacity of the wetlands in conjunction with projected runoff volumes from standard design storm frequencies and duration. Width, depth, length, and slope also depend upon desired operation and subsequent location of the spillway, as well as soil erodibility and type of grass cover. A bypass spillway that conveys storm runoff around the wetland will reduce sediment deposition within the wetlands but is likely to be more costly to construct because of the length and critical construction requirements. It will also require an additional water control structure to permit normal runoff into the wetlands while diverting excess flows through the spillway around the wetlands. Consequently, less costly methods commonly locate spillways adjacent to or in the end of a dike, thus exploiting the velocity reduction factors — vegetation roughness and impounded waters — within the wetlands system. The design must then provide an adequate cross-sectional area (width and depth) to pass peak discharge volumes from a specified rainfall event without eroding the dike.

Needless to say, the spillway should be constructed of or in erosion-resistant materials, covered with a dense stand of sod-forming grass with a 15 to 25 cm height and have a slope downstream of not more than 3 to 4 %. The spillway may simply be a low spot in the dike or a short channel around one end of the dike. In either case, it should be straight or, at most, shallowly curved with grass-covered 3:1 slopes on either side, and the slope should drain towards the downstream side of the dike. The spillway with a shallow slope should be continued fully to and meet the downstream discharge channel at a narrow angle, (i.e., the spillway should join the discharge channel at an angle of less than 20°) to reduce potential erosion on the spillway and in the discharge channel. Final grading of the spillway should ensure that depressions are not present in the spillway that would hold water and jeopardize grass cover or integrity of the dike.

Storm runoff is a function of:

1. rainfall amounts and expected frequency
2. infiltration rates of the watershed soils
3. land use and vegetative cover conditions
4. slope of the land in the watershed

The Soil Conservation Service (SCS) of the U. S. Department of Agriculture has extensive experience in monitoring and predicting runoff volumes from different storm events on small watersheds with different soils, vegetative cover, and slopes throughout the U.S. and its territories. The district office of the SCS can provide assistance in determining soil types in the targeted watershed and invaluable local information on interpretation of vegetative cover and slopes in estimating storm runoff. Local experience is also useful in interpolating rainfall amounts expected from a 24-h storm of 10, 25, 50, or 100-year frequency since regional factors (large water bodies, hills, or mountain ranges) can significantly influence local rainfall patterns. SCS specialists have considerable experience at modifying general guidelines for all aspects of designing farm ponds, including runoff estimation and spillway requirements, in most areas of the U.S. Adequate design needs for protection of a wetlands system differ little from comparable requirements for a farm pond. The typical created wetlands simply have less depth and greater surface area than most farm ponds. Before final drawings are prepared, the SCS specialists should be asked to review runoff calculations and emergency spillway dimensions to insure the design is adequate. While in the SCS office, request a copy of *Ponds — Planning, Design, Construction*, SCS Agricultural Handbook No. 590, and study the information on all aspects of pond construction.

Table 2   Runoff Curve Numbers Interrelating Soil Infiltration Capacity, Land Use, and Vegetation Cover[a]

Land use and cover	Hydrologic soil group			
	A	B	C	D
Cultivated				
Without soil and water conservation treatment	72	81	88	91
With soil and water conservation treatment	62	71	78	81
Pasture				
Poor cover	68	79	86	89
Good cover	39[b]	61	74	80
Meadow	30	58	71	78
Woods, shrub, or forest				
Thin, poor cover, no mulch	45	66	77	83
Good cover, mulch and humus	25	55	70	77

[a]   Adapted from *Ponds - Planning, Design, Construction,* SCS Handbook No. 590.
[b]   Use a runoff curve number of 60 if the table value is <60.

Predicted rainfall amounts are obtained from the *Rainfall Frequency Atlas of the United States* and the *Precipitation Frequency Atlas of the Western United States,* prepared and distributed by the U.S. Weather Bureau. These atlases include maps depicting the expected amount of rainfall during a 24-hour storm event based on 10, 25, 50 year, etc. storm frequencies. In each case, the rainfall amount is that amount expected from a storm that will, on average, occur only once in 10, 25, 50 or more years depending upon the map in question. Conversely, the 25-year storm will only have a 4% probability of occurring in any given year and the 50-year storm will have a 2% chance of occurrence in any year. Rainfall amounts in excess of the amount predicted for a 50-year storm event are only anticipated to occur, on average, once in more than 50 years, or once during the next standard category, every 100 years.

Practical construction considerations and impact severity in the event of failure suggest that wetlands in watersheds of less than 20 acres with average soils, slopes, and vegetative cover should have spillways designed to accommodate the 24-h 10-year storm event, and those in larger watersheds should be designed for a 24-h 25-year storm event. As a generality, rainfall from the 10-year storm in the eastern U.S. ranges from 9 in. near the Gulf Coast to 3 in. in northwestern North Dakota and comparable values for a 25-year storm are, respectively, 11 in. and 3.5 in. For example, values for Tennessee are approximately 5 in. for a 10-year storm, close to 6 in. for a 25-year storm, and 6.5 in. for a 50-year storm. Since rainfall amounts are not greatly different between storm categories, a conservative designer would use the next higher category to insure the adequacy of his spillway dimensions.

The SCS has developed a series of curves and equations to estimate volumes and peak discharge rates from four hydrologic soil groups with different infiltration rates, interrelated with nine land use and vegetative cover classes, as shown in Table 2. Soils in group A have high infiltration rates and those in group D have low infiltration with high runoff potentials. Runoff curve numbers obtained from Table 2 may then be multiplied by the percentage of the watershed in each category to obtain an average

Table 3. Determination of Runoff Depth[a]

	Equation constants		
Runoff curve numbers	A	B	C
CN 60	−7.698	−169.37	−23.63
CN 70	−16.75	−538.52	−33.54
CN 80	−36.89	−1948.69	−53.94
CN 90	−128.11	−18550.96	−145.54

where: $I = A + B/(C + X)$
I = runoff depth, in inches
A, B, and C = constants
X = rainfall, in inches

This expression provides a fairly good approximation for rainfall amounts between 1.5 and 12 in.

[a]Adapted from tabular data in *Ponds - Planning, Design, Construction,* SCS Handbook No. 590.

runoff curve number for that watershed. This is related to expected rainfall, in Table 3, to determine the volume of runoff from a watershed of given size using the conversion factor of 325,851 gallons per acre-foot.

For example, a 200-acre moderately sloping watershed with 110 acres in good pasture on soil group A and 90 acres in poor woods on soil group C would have an average total runoff of 7.6 million gallons after a 5-in. storm event.

$$0.55 \times 39 = 29.25$$
$$0.45 \times 77 = 34.65$$
$$= 63.90$$

From Table 3, 5 in. of rain on a 200-acre watershed with runoff curve number of 63.9 or 60 will produce

$$I = A + B/(C + X)$$
$$I = -7.698 + (-169.37/(-23.63 + X))$$
$$I = -7.698 + (-169.37/(-23.63 + 5))$$
$$I = 1.39 \text{ in. of runoff}$$

Then,

$$200 \text{ acres} \times 1.4 \text{ in.} = 280 \text{ acre-in.}$$
$$280 \text{ acre-in.} / 12 = 23.33 \text{ acre-ft}$$
$$23.33 \text{ acre-ft} \times 325,581 \text{ gal/acre-ft} = 7,595,804.73 \text{ gal}$$
$$\text{or approximately 7.6 million gal}$$

For perspective, if the new wetlands has a water surface area of 23 acres, this storm would raise the water level slightly more than 1 foot if no discharge occurred.

In addition to the total runoff, we need to estimate the peak discharge rate in order

Table 4.   Determination of Peak Rates of Discharge[a]

Runoff curve numbers	Equation constants	
	A	B
Flat slopes		
CN 60	−8.41	3.53
CN 70	−11.99	4.61
CN 80	−15.45	5.89
CN 90	−22.19	7.76
Moderate slopes		
CN 60	−16.25	6.19
CN 70	−21.86	7.92
CN 80	−29.15	9.90
CN 90	−47.28	12.97
Steep slopes		
CN 60	−28.71	10.47
CN 70	−42.92	13.55
CN 80	−55.82	16.43
CN 90	−72.06	20.13

where: $P = A + B \sqrt{X}$
P = peak discharge rate (cfs/inch of runoff)
A and B = constants
X = drainage area, in acres

This expression provides a fair approximation for 30 to 1000-acre watersheds.

[a] Adapted from graphic data in *Ponds - Planning, Design, Construction,* SCS Handbook No. 590.

to determine the dimensions for the emergency spillway as follows. From Table 4,

$$P = A + B \sqrt{X}$$

where P = peak discharge rate in cubic feet per second per inch of runoff

A and B = constants
X = drainage area in acres

For a curve number of 60 and moderately sloping watershed,

$$P = -16.25 + 6.19 \sqrt{200}$$
$$P = 71.29 \text{ cfs/in of runoff depth}$$

and Q = 71.29 cfs/in. × 1.39 in. = 99.09 ft³/s at the peak discharge rate during this storm.

For a 50-ft spillway with a 3 to 4% slope and good grass cover of 6 to 10 in. on erosion-resistant materials and maximum velocity of 5 ft/s, the dimensions of the spillway should be (see Table 5):

$$W = Q / q = 99.09 / 3 = 33.03 \text{ ft wide, and}$$
depth of flow (Hp) for a 50-ft long (L) spillway will be 1.4 ft

Table 5.   Determination of Discharge and Flow Depth[a]

Maximum velocity (V), ft/s	Discharge (q), cfs/s/ft	Flow depth (Hp) Spillway length (L), ft	
		50	100
4	2	1.2	1.4
5	3	1.4	1.6
6	4	1.6	1.8

[a] Adapted from *Ponds - Planning, Design, Construction*, SCS Handbook No. 590.

Since the minimum freeboard on the dike should be 1 ft. the emergency spillway should be built 33 ft wide and the spillway surface should be 2.4 ft below the top of the dike (deep). However, since the spillway may need to function prior to good grass cover establishment, the calculated width is an approximation and to compensate for rounding errors, a conservative design would add 10 to 20% to each dimension, or

$$33 \times 1.1 = 36 \text{ ft wide and}$$
$$1.4 \times 1.1 = 1.5 \text{ ft deep}$$

Location of the spillway is also important since it must convey excess waters around or over the dike without damaging the end, top, upstream, or downstream surface of the dike or the discharge channel. In some cases, a wingwall or short stub dike may be needed to guide stormwaters around the end of the dike and in others rock riprap may be needed on portions of the wingwall, spillway, or dike surface.

The example calculations, while developed specifically for excavated spillway estimates, also provide a fair approximation for natural spillways. In addition, they tend toward the conservative side, but most designers will do likewise since proper functioning of the emergency spillway is crucial to the security of the dikes and the continued well-being of the wetlands communities.

Preparation of Plans

Duplicate copies of the detailed site topographic maps are handy for developing initial plans. Soil types, subsurface formations, hydrologic characteristics, cultural artifacts, utility rights-of-way, biotic components, potential borrow areas, and any other factors identified in the site surveys that could be useful in selecting locations for dikes, water control structures, visitor facilities, and access can be included or prepared as overlays. Including everything on one map will likely clutter the map, so only factors relevant to locations of construction activities should be included.

With the ease of duplication today, numerous copies of the site map facilitate sketching in potential locations for various structures without having to erase and redraw every mistake. In addition, sketches of alternative locations can be prepared from which preliminary calculations of acreage needs, runoff estimates, cut and fill requirements, etc. can be developed for comparative purposes. Much of the initial work is likely to be done with basic instruments — scaled rulers, compass, or dividers — even in those instances where planners have access to sophisticated computer programs. It is simply much faster to sketch out the first rough plans. However, computer programs are now available into which the topographic survey information is fed and fairly rapid drawings can be generated directly on electronic representations of contour maps. Since elevations are included in the survey information, depicting potential depths and area of inundation is easily accomplished. A few advanced programs have the ability to create a drawing, calculate the amount of cut and fill required for specific designs, and store each drawing and calculations as a separate file so that comparative evaluations can be printed out for study and discussion purposes. Rough sketching is a bit slower with these, but much more information is available for later comparisons.

Regardless of the means employed, planners should explore all possible alternatives for developing the selected type of wetlands within each of the potential sites under evaluation. Because of the many possible interactions of project goals, site characteristics, feasibility and costs, it is not infrequent that a seemingly unusual design furnishes an optimal combination of compromises; and of course, each draft design will represent a different set of compromises since the ideal site can rarely be found.

Once initial sketches have been prepared and evaluated and a site selected, concept plans should be developed, preferably with one of the many available computer-assisted drawing programs. Though this step can also be accomplished with rudimentary drawing devices, developing the concept plans on an information management system facilitates modification, evaluation, and duplication. Numerous programs are available and some are very powerful yet inexpensive and user friendly with relatively short learning curves. Concept plans should include locations and dimensions of all structures, type and location of construction activities, and location of hazardous or cautionary areas such as utility rights-of-way, subterranean caverns, etc. With computer drawing tools, depiction of perspectives of the completed system from different viewing angles can be easily developed to assist in refining overall aesthetics. Depending upon the program(s) used and printing capabilities, the concept plan may be expanded and done in more detail to prepare construction drawings and specifications.

Construction drawings and specifications should have sufficient detail so the contractor can readily understand what is expected of him and the developer has sound basis for legal recourse if the final product does not meet contract specifications. Generally, well prepared, clear, and detailed plans and specifications will enhance the probability of the final product resembling the planned system; but investment in preparation of drawings should be proportionate to overall project complexity and cost.

Minimally, drawings and specifications must include clearing and grubbing limits, final grades, utilities, borrow areas, type of structures, dimensions, dike and structure materials, bottom permeability, valves or controls, areas or vegetation to remain undisturbed, erosion control measures, and planting requirements including sodding or seeding and mulching dikes, spillways, and other disturbed areas. A more detailed plan might include composition and type of valve or even a specific manufacturer's model number, spacing of reinforcing rod in concrete, composition and grade/thickness of piping, etc.

Ideally, construction plans will include

1. boundaries of construction activities, including clearing and grubbing limits
2. access for construction equipment and transportation corridors
3. locations of cautionary or hazardous areas
4. utility rights-of-way and contacts
5. quantities, location, and dimensions of borrow areas
6. areas or vegetation that should not be disturbed
7. erosion control measures to be taken during construction and revegetation methods during final stages
8. locations, dimensions, and materials specifications for structures
9. locations, length, top and base widths, elevations, upstream and downstream slopes, permeability and coring for dikes or berms and spillways
10. type, size, location, materials, and elevations of water control structures
11. pond bottom and side permeability specifications and methods to attain required permeabilities, including liners and liner installation if needed
12. elevations, slopes, and contours of pond bottoms and permissible tolerances
13. elevations, dimensions, composition, grades/thickness, manufacturer, and/or model for piping and valves or other water control structures
14. dimensions and specifications on lighting, switches, wiring, outlets, pumps, and other electrical facilities
15. type and method of placement of sand, gravel, rock or rock riprap
16. species, sources of supply, planting spacing, planting dates, and expected survival of wetlands vegetation
17. seeding, fertilizing, mulching and liming, or sodding of dikes, berms, spillways, and any other disturbed areas
18. provisions for on-site construction supervision
19. methods for determining permeabilities and other contract specifications
20. types, sizes, and numbers of construction equipment

Planners should check and recheck and have a competent colleague verify elevations of pool bottoms and area, dike tops, piping, and controls or valves in the drawings. Discovering during construction, or even later, that system operation will require water to flow uphill can be embarrassing and costly. Simple arithmetic errors easily result in that unusual phenomena. They should also compile a list of materials, check availability and prices, and develop cost estimates for each phase of the project to assist them in evaluating construction bids.

Cost estimates should include

1. contour mapping surveys and construction staking
2. preparation of construction drawings and specifications
3. preparation and distribution of bid invitations and advertisements
4. site preparation: clearing, grubbing, and dewatering if needed
5. categorized construction activities for major units (i.e., dikes, water controls, roadways, spillways, visitor facilities, etc.)

    a.   materials
    b.   equipment
    c.   labor
    d.   supervision
    e.   overhead percentages
6.   planting wetlands vegetation
7.   revegetating disturbed areas

Preparation for Construction

Construction drawings and specifications will become core elements in the contract for constructing the wetlands. The contract should also include construction and planting periods and any restrictions on either, type of earth-moving or other construction equipment, provisions for site access and security, completion dates and penalties or early completion bonuses, performance bonding requirements, methods for testing permeabilities, methods and dates for determining plant survival and replanting if necessary, start-up and acceptance procedures, and any other aspect that could be confused or result in a later disagreement. As with construction drawings, larger, more complex (hence, more costly) projects generally require more carefully thought out and detailed contracts. However, even in the largest projects, a certain amount of good faith and understanding between the developer and contractor is usually necessary since including every minute detail in the contract is virtually impossible. Of course, payment arrangements at the completion of certain phases or at final completion and acceptance should be included.

Once the contract has been developed, the invitation to bid is abstracted from the contract and advertised as required of public agencies and/or through local news media and contractors association newsletters. Invitations should be sent to potential contractors known to planners or identified by the local highway department, SCS office, or trade associations. Advertisements should contain a brief description of the project, bid deadlines and procedures, expected completion dates, information on obtaining the invitation to bid, and a contact for further information. Invitations to bid should include a conceptual drawing and general specifications, unusual requirements, deadlines, construction dates, penalties, and bidding forms. Breaking down the project into discrete components and requiring bids for each component on bidding forms will facilitate evaluation and comparison of bids received. In large or complex projects, bid invitations often include complete construction drawings and the proposed contract.

Since wetlands construction is likely to be unusual for most prospective contractors, the invitation to bid and advertisements should include a time, date, and location for a pre-bid site inspection and conference. Although wetlands construction is essentially the same as building a shallow lagoon, many potential bidders will have reservations and will likely increase bid amounts, in some cases astronomically, as insurance. A thorough explanation of project goals, construction expectations, and a site walk-over will allay many fears and promote realistic bidding.

Prior to the pre-bid conference, the information on construction plans must be transposed to the site through staking so that potential contractors can readily determine the expected work. Hopefully, basic stakes and elevations are still present from contour mapping and available for a baseline. Placing construction stakes locates individual construction activities and delineates the elevations, grades, and lines specified in the plans. Boundaries, depth of cut, elevation of fill, slope angle and position, location and elevation of piping and structures, and all of the other information on construction plans should be transferred to the site through clear and ample stakes. Staking must clearly show contractors what, where, and how much is to be done to facilitate bid preparation and to insure the finished product adheres to construction plans and specifications.

Area staking consists of marking access, boundaries of the construction area, clearing and grubbing limits, the proposed water level, and borrow areas. In addition to boundary stakes, borrow areas must have cut stakes to indicate excavation depths to insure that proper fill materials are obtained and not unsuitable materials that may underlie the borrow area.

Dikes are located by placing stakes along the centerline and fill and slope stakes upstream and downstream at the point of intersection with the ground surface. Spillways are staked along the centerline with cut and slope stakes at the intersections with ground surface. Locations, dimensions, and elevations of water control structures, visitor facilities, roadways, utilities, and other structures should also be clearly marked so that construction workers can readily determine project components. Planners also benefit from seeing the project staked out on site. More than one set of plans has been revised after the site was staked for construction.

Despite careful and thorough site inspection and evaluation, planning and preparation of drawings and specifications, some eventuality is likely to be overlooked. Consequently, completion of invitations to bid, construction plans, a contract and construction staking will only mark the completion of the major portion of project planning. Quite likely, construction conditions and activities will require revisions in the original plan, which is the reason that on-site supervision by competent personnel and/or ready access to project planners is critical during the construction phase.

Planning activities grade into construction activities as planning, construction staking, the pre-bid conference, and contract award(s) follow one another. Probable delays before equipment actually moves on site include minimum advertisement periods for public agencies, contractor's scheduling on other jobs, and inclement weather. Since the final construction activity is planting vegetation or improving growing condition for native vegetation, and both should be accomplished during spring or early summer, project schedules and construction time frames must be carefully established and clearly understood by planners and contractors. Generally, construction is most rapid in dry seasons, typically late summer and fall, in which case planting should be delayed until the following spring. In the interim, flooding the system may be useful to check grades, elevations, sealing and operation of piping and water control structures, electrical accessories, and other facilities. Identified errors or problems can then be corrected before vegetation is placed in the pools the following spring. If bentonite was used, the pools should be flooded until just prior to planting to avoid cracks and leaks.

Construction activities include clearing and grubbing, excavation, grading, transporting and placing fill, compacting, placing sand, gravel, or rock riprap, installing liners, placing and tilling in sealing substances, disposing of waste or excess fill, building or installing water control structures and piping, installing electrical facilities and other utilities, planting wetlands vegetation, seeding, and mulching or sodding disturbed areas,

Schedules for various phases of construction should be agreed to prior to any activities so that moving equipment in and starting work do not interfere with other planned activities on site and nearby site disturbance is kept to a minimum. Dry weather, generally late summer and fall, is typically optimal for construction and the job may be finished quickly. During wet seasons, contractors may only be able to work for 2 to 3 days and then be idle for days or weeks, dragging out the period when the site is disturbed; erosion may be high and construction activities disrupt other site work. Scheduling should also include consideration of neighboring landowners and land uses to cause as little disturbance to them as possible. Even though it may seem advantageous to have heavy equipment working at night to complete the job, disrupting the neighbors relaxation or sleep will set the seeds for future problems.

Prior to construction planners may need to

1. investigate and, if necessary, divert or pump water from the site
2. mark any trees with flagging that should be left or limbed or anything other than clearing
3. identify locations of silt barriers (with fence or straw bales) needed, and insure that contractors comply with regulations and any special requirements
4. Discuss equipment types and numbers with contractors to insure expeditious work activities; improperly sized equipment will increase construction time and costs and may inhibit accurate construction. For example, top width of the dikes will be at least as wide as the bulldozer blade because it is difficult for the operator to build anything less. If the site is wet, at least two dozers or other machines should be used since one will often be needed to extract the other after it mires down. Obtaining a unit off-site each time the dozer is stuck will cause needless delays. If the site is very wet, a dragline using supporting mats is generally more efficient than dozers and scrapers or backhoes.

The system planner or a construction supervisor familiar with all aspects of the project (including design objectives and management plans) should be on-site or at least monitor activities on a daily basis. Invariably, some aspect has been overlooked or is not anticipated, and modifications will be necessary (see Figure 1). If someone knowledgable is not available, contractors may stop work activities until advised, or independently modify the plans. In the first case, valuable time will be lost; but in the second, major errors could occur that could be costly to correct and contractors may be unwilling to absorb the extra costs.

Specific construction activities vary substantially with site conditions, type of wetlands, and construction equipment and work force. However, construction supervision is primarily insuring that the contractor adheres to construction drawings and specifications and developing and agreeing to modifications as necessary to accommodate unforeseen circumstances. Modifications should be approved by the system planner or someone familiar with the technical specifications, future management plans, and requirements before adoption. Though construction supervision is as varied as different projects, general guidelines and precautions are applicable to a variety of wetlands construction activities.

Topsoil should be removed and stockpiled for later use during the early stages of clearing. If a wetlands soil is removed for later use, store it underwater to avoid oxidizing and releasing bound metals or other substances that could detrimentally impact the new system if they were re-dissolved. All permeable soil materials, organic matter, rocks, trash, or debris should be removed in preparing a solid, impermeable foundation for dikes and other structures. Stream beds must be widened and deepened, and all stones, gravel, or sand removed to the clay foundation so that fill material will bond properly. Natural holes or holes caused by clearing and grubbing should be cleaned out and filled with suitable fill material. In each instance, the objective is to insure that clay fill materials will abut and bond to a clay foundation to insure continuity of the impervious materials.

Waste or spoil materials that will not be needed for fill should be placed in boundary areas, sloped and contoured to blend into the surroundings, and stabilized by seeding or sodding. Excess earth could be used to provide knoll overlooks, nesting islands, wider parking areas and visitor stops, visual and/or auditory screening, or any other imaginative use that will not interfere with the principal functions of the new wetlands.

Construction should always follow contract specification as closely as possible. In

Figure 1.   Extensive stands of bulrush in a created marsh in the Dakotas provides important nesting habitat and shelter for a variety of wetland wildlife.

many wetlands, slight deviations may not be critical; but if the system will be used for wastewater treatment, grading must meet the specifications in the plan within described tolerances to achieve proper functioning of the system. Out-of-tolerance lateral bed slopes may not only cause ponding, but are likely to cause channeling or short-circuiting that reduces the effective treatment area in the cell and depresses performance. Similarly, improper grading along the cell length could make it impossible to set and maintain proper water depths, as well as causing channeling and short-circuiting. Obviously, inability to manage water depths would severely retard establishing or managing the desired plant community and retard functioning of any type of wetlands.

Specifications on dike materials and dimensions are established to reduce or eliminate leakage or seepage and to insure dike integrity under expected water pressures. Failure to follow specifications on materials, top width, base width, or slopes could result in weak or leaky dikes or highly erodible slopes that may be difficult to stabilize. Installing the clay core and culverts or controls in the dikes must also be done carefully, with proper fill compacting to prevent seepage.

Meeting permeability specifications is important in all wetlands. Be careful that contractors do not excavate deeper than planned and penetrate an impermeable layer into a permeable layer. Permeability testing should be agreed upon prior to construction, and frequent testing is necessary during construction. Compacting in situ or fill material must be done with proper equipment and only when moisture conditions are satisfactory. If necessary, sprinklers may be used to achieve proper soil moisture conditions before and during compacting. Bentonite or soda ash blankets must be carefully installed and tilled into the bottom following specifications in the designs to insure proper functioning. If synthetic liners are used, installation must follow manufacturers' instructions for bed material, sealing (liner-to-liner and liner-to-piping and controls), and insulating material above the liner. Construction equipment and workers can easily puncture synthetic liners and clay blankets unless special precautions are communicated and followed. Once final grading and compacting has been accomplished or a liner installed, equipment should not be permitted on the cell bottom or dike sides, and foot traffic should be minimized.

If parent or fill materials are very fine clays that easily powder after disturbance, plan to flood the system and dewater it before final grading. Preliminary flooding is also the most simple method to check grade and structure elevations throughout the system. Final grading before flooding may need to be repeated since it is not unusual for settling to cause elevation differences of 15 to 30 cm in cell beds and occasionally on dikes.

Installing water control structures may be as simple as laying PVC pipes, or as complex as building forms and pouring concrete in place. In either case, correct elevations and adequate support are critical to future management ability. Correct proportions of cement and sand and approved temperatures are important in achieving design strength for concrete structures. Appropriate adhesives, bonding techniques, and curing temperatures are required to insure correct joining of plastic piping. As always, specified elevations for each and every segment of pipe or portion of a concrete structure are especially important. Generally, wetlands system operation is dependent on gravity flow that is only possible if design elevations are achieved in all aspects of the system.

In as much as vegetation planting requirements vary considerably with region, site conditions, species, and planting season, detailed information is provided in the following chapter. In general, planting supervision, as with other construction activities, is principally insuring that contractors adhere to contract specifications so that design objectives will be achieved.

Optimal planting conditions for cut materials or seed are created by shallow flooding, followed by dewatering but not complete drying — leaving soft, moist soil conditions. Common pitfalls include improperly storing planting materials — too dry or too warm or too long — careless handling and poor orientation. Planting stock should not be dug more than 2 days before planting and should be stored and transported in a cool, dark, humid environment. Damaged stock may perish or take longer than normal to begin growth and wetland plant roots are as sensitive to damage as any garden plant. Rows must run perpendicular to the direction of flow to improve coverage and reduce channeling, even though it may be easier to operate equipment up and down the long axis of each cell. After planting is completed, flood the cell with 1 to 2.5 cm of water, but insure that water depths do not overtop cut stalks or the new plantings may die. As new growth begins, water levels may be slowly raised, but should not overtop the new growth.

Hand or natural seeding is less expensive, but much less reliable for starting the new plant community. Germination rates of many wetland plant seeds are less than 5%, so large quantities must be collected and distributed. Whether hand or natural seeding is used, the pond should be shallow flooded in late winter and early spring and dewatered at the onset of warm weather to establish warm, moist mud conditions. Careful monitoring and regulation of water levels at or just below the pond bottom is important in order to maintain the proper soil moisture conditions for germination and sprouting. After the new growth has reached 10 to 12 cm, water levels should be raised to between 2 and 4 cm above the substrate to inhibit or kill terrestrial species, but should not overtop wetlands plants.

Trees or shrubs along the perimeters should be installed after a good grass cover has become established in the event that re-grading and re-seeding should become necessary because of erosion or other problems. Generally, trees or shrubs should not be planted on dikes unless the dike is very large and root growth is not likely to impugn the integrity of the dike. Similarly, fencing for livestock or security should not be installed until after disturbed areas have become well covered with stabilizing vegetation.

All control structures, piping, wiring, pumps and other electrical facilities, seals, water levels, and flow distribution should be tested for proper operation or elevations before formal acceptance. Plant survival can be estimated by observing the percentage of plants with new and vigorous growth evident. Since some aspects — settling, subsidence, leaks, or seepage — may not become apparent until some time after initial operation, planners should establish a test or start-up period with the contractor and arrangements that may be needed to correct problems. Elevation variances will largely become apparent with initial flooding, as will problems with piping and controls. However, seeps or leaks may be undetected for fairly long periods, and plants showing good growth initially may die later from any number of factors. Since some of these are controlled by project personnel (i.e., water depths) planners should obtain agreements with contractors on remedial actions and responsibilities.

Any erosion on dikes, spillways, around control structures, or in the upper ends of cells should be immediately filled with clay or soil materials, compacted, and re-seeded or re-sodded as necessary. Before the vegetation provides a protective screen, wave action may erode dikes or earthen fills and temporary log booms or slat fencing may be needed. If seepage or leaks are detected, water levels should be lowered immediately. Depending on the location, adding additional sealing material, bentonite, or soda ash may correct the problem. However, leaks around control structures or at the base of a dike will require excavation of fill materials and replacement with compacted, impervious clays. If a synthetic liner was used, additional panels may need to be added and securely bonded to previous panels.

In summary, proper construction supervision is principally having a knowledgable person ensure that construction drawings and specifications are followed precisely and developing and documenting modifications as necessary. Since experience levels and work quality may vary substantially between contractors, time and effort required for supervision will also vary, but close attention to details is always a prudent investment. Since few contracts fully describe every component, a considerable amount of faith and trust between planners and contractors is essential. However, even the best contractors may misunderstand or misinterpret drawings and specifications, or more likely, encounter unforeseen situations, and readily available advice or directions will save time and expense and possibly, future management capabilities. Supervision also includes frequent and thorough inspections of all aspects of the project during start-up and immediate attention to any problems. In as much as some problems may not be identified until after the contractor has moved off site, good working relationships between the construction supervisor and the contractor are important in implementing remedial actions before minor faults become major failures.

SELECTION OF SPECIES

The diversity and complexity of natural wetlands are prinicipally the result of interractions of three important factors: (1) hydrology, (2) substrate, and (3) vegetation. The first two strongly influence the vegetation, as do climate and proximity to other wetlands. Since restoring or creating wetlands is dependent on duplicating these factors and their interactions, an understanding of the ways they influence vegetaion is important to proper selection of species, planting or establishment methods, and operating conditions.

Hydrology

Water depths, frequency and duration of flooding, and water chemistry are the most important factors determining the survival and growth of plants in a wetlands system. Depth influences gas exchange between the substrate and the atmosphere, decreasing oxygen contents below deeper waters. Depth also restricts light penetration, even in clear waters, though highly turbid waters are unlikely to permit adequate light penetration to support photosynthesis much below the surface. The vegetation zonation characteristic of most wetlands is largely due to the influence of water depth because certain species are adapted to or require certain depths, whereas others prefer different depths. In fact, a seasoned observer can judge water depths and bottom types in different wetlands systems simply by noting the plant species present in each area. Much of the diversity and spatial heterogeneity of natural wetlands systems is the result of different elevations and, consequently, water depths within the system.

Frequency, duration, and seasonality of flooding also strongly influence species that should be used for developing different wetlands or different areas within the wetlands. Different wetland plants withstand various degrees of inundation depending on when and for how long the flooding occurs. Most shrub and tree species and some emergent plants need a period of lower water levels or very limited flooding during the growing season, but can withstand prolonged inundation in the dormant season, generally fall and winter. In dormancy, plant oxygen needs are reduced and long and/or deep flooding may have little impact. However, oxygen limitations during the active growing period may cause stress and eventually mortality in the same plants. Some swamp trees can endure fairly long periods of shallow flooding, but may be damaged by deep flooding; whereas others are vulnerable to any inundation beyond 4 to 5 days during the growing season.

In natural wetlands, various grasses occupy the highest regions that either have very shallow flooding (2 to 5 cm) or limited seasonal flooding (see Figure 1). Maidencane and reed canary grass occur in the deeper water portions of this wet meadow zone slowly grading into sedges, *Iris,* rushes (*Juncus*), woolgrass (*Scirpus cyperinus*), and spikerushes (*Eleocharis*) with increasing depths. These are, in turn, replaced by arrowhead, sweetflag, pickerelweed, and water plantains further inward. Cattail and giant reed often intergrade with the last zone, but may extend out to depths of 15 to 30 cm with the large bulrushes (*Scirpus validus*), often occurring in 30 to 50-cm depths. Few emergents colonize deeper waters. At this point, submergents (pondweeds, coontail, or tapegrass), become dominant if water chemistry and clarity are suitable. In the deepest regions (1 m and more) a few pondweeds (e.g., *P. robbinsii*) may be able to grow if clarity is high; but in turbid or colored waters, only rooted floating species, spatterdock, and water lilies persist. Similar depth and flood duration-dependent zonation occurs in wooded wetlands with characteristic tree-shrub complexes present at various elevations and in various flooding regimes. However, only a few (cypress, black willow, water hickory, water elm, water tupelo, water locust, swamp privet, green ash, Nuttall's and overcup oaks) can survive long-term inundation (i.e., flooding during the growing season) for more than 1 year. A few (cypress, tupelo, willow, and overcup oak) grow in standing water for many years, similar to emergent herbaceous species. However, most wetlands trees and shrubs are only adapted to endure short flooding periods during the growing season, perhaps long periods during fall and winter, but not permanent inundation as are cattail or bulrush.

A selection of herbaceous and woody species grouped by ranges or duration of flood endurance is presented in Tables 1 through 6. This list is not all inclusive, nor will all species survive in all regions of the country or all types of wetlands. Planners should use the list as a guide for comparison with herbs and trees endemic to their region to select species for planting in different depths of the new wetlands. Obviously, it would be just as inappropriate to attempt to establish cypress in northern Manitoba as it would to plant black spruce or tamarack in Florida.

To repeat, the best guidelines for selecting plant species for the new wetlands will come from observing a similar wetlands in the vicinity. Bear in mind that new plantings or seedlings will be less able to survive extremes than will mature individuals. Seedlings of tolerant species planted among mature representatives of less-tolerant species have been killed by prolonged flooding during winter that stressed, but did not kill, the less-tolerant species (see Figure 2).

The categories in Tables 1 through 6 (flood tolerance) are defined by maximum flooding tolerance of mature individuals of native species. Nursery stock with different genotypes may be much less tolerant of prolonged or deep flooding. Conversely, shrubs and trees placed in a wet category will usually thrive in a dryer regime. For example, cypress was planted on reservoir shorelines and in sinkholes in east Tennessee, far out of its normal range. The shoreline stands rarely are flooded and the sinkhole

Figure 1.   Upland grasses occupy dry sites in the foreground, with transition species near water, emergents in shallow reaches, and submergents in the deep zones.

Figure 2.   Cordgrass occurs in the transition zone, the far right grading into arrowhead, and then a dense mat of American pondweed in deeper waters in the background.

Table 1.   Environmental Requirements of Selected Herbaceous Wetlands Plants

Plant type	Soil	pH	Salinity (ppt)	Depth (cm >, < water surface)
Spartina cordgrasses	Sandy loam	5—8	10—30	20 to −50
Calamogrostis reedgrass	Silt	5—8	0—10	20 to −20
Phalaris reed canary	Silt loam	6—7.5	0—0.4	10 to −10
Juncus rushes	Organic	5—8	0—30	10 to −10
Cyperus nutsedges	Organic soil, clay	3—8	0—0.4	10 to −30
Carex sedges	Organic soil, clay	5—7.5	0—0.4	10 to −50
Typha cattail	Organic soil, silt clay	5—710	0—25	10 to −70
Phragmites reeds	Silt, sand, clay, organic soil	3—8	0—30	30 to −150
Scirpus bulrushes	Organic soil, clay	4—9	0—32	10 to −100
Potamogeton pondweeds	Organic soil, silt clay	4—10	0—20	−5 to −300
Vallisneria tapegrass	Silt clay	5—8	0—10	−5 to −100
Ruppia widgeongrass	Silt clay, organic	5—10	0—25	−5 to −100
Nuphar spatterdock	Organic soil, silt	3—8	0—5	−50 to −200

stands are only flooded in wet years, but robust populations have existed for 20 to 40 years. Relatively complete canopy cover resulting from closely spaced plantings has virtually eliminated invasion by competitive species. At the other extreme, willow oak is frequently planted as a shade tree or ornamental on city streets, even in hilly, well-drained areas, and some of the largest examples occur in occasionally flooded city parks in the Piedmont and Coastal Plain.

Physical and chemical parameters of water also affect survival and growth of wetland plant species. These include water clarity, pH, dissolved nutrients, salt concentration, flow velocity, and dissolved oxygen. For submerged species, water clarity is important since light penetration is reduced in turbid or stained waters, thereby limiting photosynthesis. Under these conditions, rooted aquatics with floating leaves (*Nymphaea, Nuphar, Brasenia, Potamogeton nodosus,* and other floating leaved pondweeds) can overcome physical light penetration limitations. In clear but fast flowing waters, species with filiform leaves and extensive root systems (*Vallisneria* and *P. pectinatus*) are resistant to current disturbance often growing on rock strata in fast flowing rivers.

Salt concentrations of the substrate or in the water column strongly influence plant survival by impacting osmotic balance, the determining factor in direction of passage of water and dissolved materials across cell membranes. Water tends to move towards

Table 2.   Characteristics of Selected Herbaceous Wetlands Plants

Scientific name	Common name	Water depth
Bidens spp.	Beggarticks	Trans
Echinochloa crusgalli	Barnyard grass	Trans
Hymenocallis spp.	Spider lily	Trans
Lysimachia spp.	Loosestrife	Trans
Hordeum jubatum	Foxtail barley	Trans
Sesbania spp.	Hemp	Trans
Polygonum lapathifolium	Pale smartweed	Trans
Iris fulva	Red iris	Trans
Osmunda cinnamomea	Cinnamon fern	Trans
Osmunda regalis	Royal fern	Trans
Setaria spp.	Foxtail grass	Trans
Spartina pectinata	Prairie cordgrass	Trans
Panicum virgatum	Switchgrass	Trans
Calamagrostis inexpansa	Reedgrass	Shall
Asclepias incarnata	Swamp milkweed	Trans
Distichlis spicata	Saltgrass	Trans
Alopecurus arundinaceus	Foxtail	Trans
Scolochloa festucacea	Marshgrass	Trans
Lysichitum americanum	Yellow skunk cabbage	Trans
Symplocarpus foetidus	Skunk cabbage	Trans
Hibiscus moscheutos	Swamp rose mallow	Trans
Hibiscus militaris	Halbeard-leaved R. mallow	Trans
Leersia oryzoides	Rice cutgrass	Shall
Juncus effusus	Soft rush	Shall
Saururus cernuus	Lizardtail	Shall
Carex spp.	Sedge	Shall
Eriophorum polystachion	Cotton grass	Shall
Cyperus spp.	Chufa, sedge	Shall
Iris virginicus	Blue iris	Shall
Iris pseudacorus	Yellow iris	Shall
Panicum hemitomon	Maidencane	Shall
Dulichium arundinaceum	Three-way sedge	Shall
Beckmannia syzigachne	Sloughgrass	Shall
Panicum agrostoides	Panic grass	Shall
Scirpus cyperinus	Woolgrass	Shall
Habenaria spp.	Swamp orchids	Shall
Cypripedium spp.	Lady's slipper	Shall
Thelypteris palustris	Marsh fern	Shall
Paspalum spp.	Knotgrass	Shall
Hydrocotyle umbellata	Water pennywort	Shall
Caltha leptosepala	Marsh marigold	Shall
Isoetes spp.	Quillwort	Shall
Phalaris arundinacea	Reed canarygrass	Shall
Sarracenia spp.	Pitcherplant	Shall
Jussiaea repens	Water primrose	Shall
Justicia americana	Water willow	Shall
Polygonum coccineum	Swamp smartweed	Shall
Polygonum pensylvanicum	Pennsylvania smartweed	Shall
Polygonum amphibium	Water smartweed	Mid
Cladium jamaicensis	Sawgrass	Mid
Acorus calamus	Sweet flag	Mid
Calla palustris	Water arum	Mid
Sparganium eurycarpum	Burreed	Mid
Zizania aquatica	Wild rice	Mid
Eleocharis spp.	Spikerush	Mid
Alisma spp.	Water plantain	Mid

Table 2.   Characteristics of Selected Herbaceous Wetlands Plants (Continued)

Scientific name	Common name	Water depth
Scirpus americanus	Three-square	Mid
Typha latifolia	Wide-leaved cattail	Mid
Typha angustifolia	Narrow-leaved cattail	Mid
Scirpus fluviatilis	River bulrush	Mid
Sagittaria latifolia	Arrowhead	Mid
Pontederia cordata	Pickerelweed	Mid
Glyceria spp.	Mannagrass	Mid
Nasturtium officinale	Watercress	Mid
Limnobium spongia	Frogbit	Mid
Peltandra cordata	Arrow arum	Mid
Menyanthes trifoliata	Buck bean	Mid
Vaccinium macrocarpon	Cranberry	Mid
Potamogeton pectinatus	Sago pondweed	Deep
Vallisneria americana	Tapegrass	Deep
Ranunculus flabellaris	Yellow water buttercup	Deep
Ranunculus aquatilis	White water buttercup	Deep
Callitriche spp.	Water starwort	Deep
Phragmites australis	Giant reed	Deep
Scirpus validus	Bulrush	Deep
Ruppia maritima	Widgeongrass	Deep
Ceratophyllum demersum	Coontail	Deep
Myriophyllum	Milfoil	Deep
Utricularia spp.	Bladderwort	Deep
Elodea	Water weed	Deep
Nymphaea odorata	Fragrant white lily	Deep
Nuphar luteum	Spatterdock	Deep
Brasenia schrebrri	Water shield	Deep
Nelumbo lutea	Water lotus	Deep
Nymphoides aquatica		Deep

Floating

Lemna spp.	Duckweed	
Azolla spp.	Water fern	
Wolffiella spp.	Wolffiella	
Wolffia spp.	Watermeal	
Spirodela spp.	Giant duckweed	
Sphagnum spp.	Sphagnum moss	

Note: Transitional = seasonally flooded; Shallow = seasonally flooded to permanently flooded to 15 cm; Mid = 15 to 50-cm water depths; Deep = 50 to 200-cm water depths.

the region of higher salt concentrations and non-adapted plants in high salt environments are dessicated by saline waters or unable to take up water and nutrients from saline soils. In alkali wetlands of the West or in coastal regions with brackish waters, cordgrass (*Spartina alterniflora*), salt grass (*Distichlis spicata*), wigeongrass (*Ruppia maritima*) or a species of *Salicornia* should be planted. Sago pondweed (*Potamogeton pectinatus*), *Zanichellia*, *Triglochin*, alkali grass (*Puccinellia*), *Eleocharis parvula*, *Eleocharis rostellata*, *Juncus acutus*, saltbush (*Atriplex* spp.), *Sueda* spp., and some forms of bulrush (previously *Scirpus acutus* and *S. fluviatilis*) are also tolerant of moderately to strongly saline conditions.

Table 3.   Woody Wetlands Plants that Will Tolerate Flooding for More than 1 Year

Scientific name	Common name
Taxodium distichum	Bald cypress
Salix nigra	Black willow
Carya aquatica	Water hickory
Planera aquatica	Water elm
Nyssa aquatica	Water tupelo
Gleditsia aquatica	Water locust
Forestiera acuminata	Swamp privet
Fraxinus pennsylvanica	Green ash
Quercus nuttallii	Nuttall's oak
Quercus lyrata	Overcup oak
Ilex decidua	Deciduous holly
Cephalanthus occidentalis	Buttonbush
Carya illinoensis	Pecan
Pinus serotina	Pond pine
Cornus stolonifera	Red-osier dogwood
Salix lasiandra	Pacific willow
Salix exigua	Narrow-leaf willow
Salix hookeriana	Hooker willow
Campsis radicans	Trumpet vine
Diospyros virginiana	Persimmon
Rosa palustris	Swamp rose

Saline waters usually have moderate to high pH levels; and the species suggested above are generally appropriate. Low-pH waters (bogs) impose an entirely different set of constraints, although pH is often less directly influential than are the associated parameters. Bogs also tend to have fairly stable water levels (little disturbance), low nutrient contents, and high organic contents, though most materials are only partially decomposed, and dark or stained waters that restrict penetration of sunlight. At extremes, low pH may directly affect plants, though some seem to adapt to ranges as low as 3.5 S. U.; for example, certain colonies of cattail (*T. latifolia*) and, of course, a number of mosses. However, many acidophilic species have adaptations that overcome nutrient or low light constraints rather than obvious adaptations to high hydrogen ion concentrations. For example, many insectivorous plants occur in bogs, obtaining their nitrogen and other scarce nutrients from insects rather than the impoverished substrate or waters. Water lilies (*Nymphaea*), spatterdock (*Nuphar*), and lotus (*Nelumbo*) have large root stocks capable of storing and supplying energy to sustain growth to reach the surface where floating leaves have abundant light energy for photosynthesis.

Typical bog trees include black spruce (*Picea mariana*), tamarack (*Larix laricina*), and balsam (*Abies*), while acidophilic shrubs include representatives from *Spirea, Vaccinium, Kalmia, Chamaedaphne, Magnolia, Ledum, Persea, Alnus, Ilex, Leucothoe,* and *Rhododendron*. Grasses often found in bogs include *Calamagrostis, Deschampsia, Muhlenbergia, Calamogrostis, Eriophorum,* and *Glyceria*. Emergents in bogs are dominated by mosses, and a few species of *Juncus, Eleocharis, Cyperus, Dulichium arundinaceum, Rhynchospora*, many sedges (*Carex*), *Peltandra,* and species of herbs from the genera of *Lachnocaulon, Helonias, Cypripedium, Habenaria, Pogonia, Spiranthes, Caltha, Sarracenia, Drosera, Parnassia, Polygala, Pin-*

Table 4.   Woody Wetlands Plants that Will Tolerate Flooding for One Growing Season

Scientific name	Common name
Kalmia polifolia	Bog laurel
Ledum groenlandicum	Labrador tea
Sambucus callicarpa	Elder
Spirea douglasii	Hardhack
Ledum groenlandicum	Labrador tea
Picea mariana	Black spruce
Larix laricina	Tamarack
Chamaecyparis thyoides	Atlantic white cedar
Vaccinium uliginosum	Blueberry
Thuja occidentalis	Northern white cedar
Populus heterophylla	Swamp cottonwood
Tsuga heterophylla	Western hemlock
Picea sitchensis	Sitka spruce
Toxicodendron vernix	Poison sumac
Magnolia virginiana	Sweetbay
Persea borbonia	Redbay
Rhus glabra	Smooth sumac
Populus fremontii	Fremont poplar
Quercus lobata	Valley oak
Salix piperi	Dune willow
Liquidambar styraciflua	Sweetgum
Populus deltoides	Cottonwood
Quercus imbricaria	Shingle oak
Quercus palustris	Pin oak
Diosphyros virginiana	Persimmon
Fraxinus americana	White ash
Acer rubrum	Red maple
Celtis laevigata	Sugarberry
Celtis occidentalis	Hackberry
Alnus glutinosa	Black alder

guicula, Calla, Drosera, Darlingtonia, Hypericum, and *Viola.* Submergents consist largely of *Utricularia* and *Menyanthes,* while *Nuphar* occurs in the floating leaved niche.

Substrate

Most common substrates are suitable for wetland establishment, but fertile loam soils are best for wetlands as for other types of plant production. Sandy loam soils are soft and friable, allowing for easy rhizome and root penetration; but heavy clay soils may restrict root and rhizome penetration. Inadequate or excessive nutrient content may limit growth and development, depending on the nutrient content of inflowing waters. If topsoil is replaced or a layer of good soil is placed above the impermeable line, soil amendments are usually not necessary because wetland plants thrive in a broad range of soil types. Occasionally, fertilizers or liming may be beneficial if the substrate is poor in nutrients or acidic. Nutrients in natural wetlands soils often become immobilized under reducing conditions in the substrate. Dewatering the substrate and oxidizing enclosed nutrients makes them readily available to plants, causing vigorous

Table 5.   Woody Wetlands Plants that Will Tolerate Flooding for Less than 30 Days During the Growing Season

Scientific name	Common name
Populus grandidentata	Bigtooth aspen
Tilia americana	Basswood
Carpinus caroliniana	Ironwood
Acer negundo	Box elder
Smilax spp.	Greenbrier
Rhus radicans	Poison ivy
Vitis riparia	Wild grape
Parthenocissus quinquefolia	Virginia creeper
Cornus stolonifera	Red stem dogwood
Quercus phellos	Willow oak
Quercus nigra	Water oak
Quercus bicolor	Swamp white oak
Quercus falcata	Spanish oak
Quercus macrocarpa	Bur oak
Nyssa sylvatica	Black gum
Platanus occidentalis	Sycamore
Betula nigra	River birch
Ilex opaca	American holly
Crataegus mollis	Hawthorn
Acer negundo	Box elder
Acer saccharinum	Silver maple
Alnus rugosa	Hazel alder
Gleditsia triacanthos	Honey locust
Ulmus americana	American elm
Ulmus alata	Winged elm

growth and development of the plant community. Macronutrients are rarely limiting in wetlands soils, but obtaining samples for laboratory analysis of soil constituents in a created system is always a good practice prior to selecting species for planting.

Sandy loam and clay loam soils normally have adequate nutrients, provide good water and gas circulation, and have moderate texture to support the new plants and to permit root or rhizome penetration. Peaty soils are generally nutrient deficient and their soft, loose texture allows new plantings to float out or fall over. However, peaty soils would be preferred if the project will develop a bog since the organic acids will depress soil pH and mats of *Sphagnum* or other mosses may be laid on top of the peat and anchored to the parent material below. Spent mushroom compost and various manures have been layered in newly constructed wetlands by developers attempting to imitate the peaty organic soils formed by many wetlands, but the benefits are doubtful.

Clays and gravels may be so dense or hard that they inhibit root penetration, they may lack nutrients found in topsoil, or they may be impermeable to water needed by roots. Plants placed in heavy clay soils may never spread because their roots are unable to extend beyond the original planting hole or their rhizomes cannot penetrate the heavy clays and initiate new shoots. Sands and gravels dry rapidly and if the water level is lowered below the level of the roots, the plants may die from dessication.

Most wetland plants have broad tolerances for normal levels of most soil nutrients, but a few are restricted to acid soils (typical bog species), some thrive in alkaline, calcium rich soils (*Chara* spp. and *Potamogeton pectinatus*), and others tolerate high

Table 6.   Woody Wetlands Plants that Will Tolerate Flooding for Less than 5 Days During the Growing Season

Scientific name	Common name
Betula populifolia	White birch
Betula alleghaniensis	Yellow birch
Betula payrifera	Paper birch
Acer saccharum	Sugar maple
Populus tremuloides	Quaking aspen
Tsuga canadensis	Eastern hemlock
Picea rubens	Red spruce
Fagus grandifolia	American beech
Carya tomentosa	Mockernut hickory
Carya ovata	Shagbark hickory
Carya cordiformis	Bitternut hickory
Carya lacinosa	Shellbark hickory
Cornus florida	Flowering dogwood
Gymnocladus dioica	Kentucky coffee tree
Juglans nigra	Black walnut
Prunus serotina	Black cherry
Quercus velutina	Black oak
Quercus alba	White oak
Quercus rubra	Red oak
Quercus shumardii	Shumard oak
Sassafras albidum	Sassafras

calcium levels (*Scirpus validus, Pontederia cordata, Ruppia maritima,* and *Salicornia* spp.). Most species that occur in brackish coastal waters or waters of the Intermountain West tolerate relatively high salt levels, but many that are found in natural wetlands of the East are intolerant of salts in the substrate or water column.

Developers often use fertilizers at the initiation of a new system, with little basis from soil analysis or from plant requirements. Since fertilizer and lime are frequently used in terrestrial planting situations, many workers presume that adding nutrients and/ or modifying the pH will benefit new wetlands systems. In a few instances, soil amendments doubtless have improved growing conditions for new plantings; however, in many cases, the nutrients were quickly taken up by algae and little was available for macrophytes because typical formulations of water-soluble fertilizers coupled with broadcast application were used. If fertilization is deemed necessary, side dressing with a poorly soluble, time-release fertilizer such as "Osmocote" or "Magamp" will obtain the best results. Both are granular, slow-release formulations that should be applied to the substrate prior to flooding. Various tablet formulations may be used if the area cannot be dewatered or if fertilizer is to be applied after planting.

SPECIES CHARACTERISTICS

Since the plants in wetlands systems provide the basis for animal life as well as conducting important hydrologic buffering and water purification functions, selection of species appropriate to project goals is important. Generally, marsh emergents (*Typha, Scirpus, Phragmites,* etc.) produce limited foods for vertebrates, though most support abundant insect populations (see Figure 3). Muskrats and beaver feed on tubers and rhizomes and a few waterfowl use leaves, stems, or tubers of *Sagittaria, Alisma, Acorus, Pontederia, Sparganium, Cyperus,* and some other types. However, emergents provide cover and protection from the weather and predators that is critical to many wildlife species. In many cases, cover is simply a dense stand of vegetation to swim or run into to avoid predators or to get out of a cold wind, but muskrats use emergents for constructing shelters (muskrat houses) and many birds construct floating or suspended nests of leaves and stems. Though generally marsh emergents are needed for cover, some species of knotweeds (*Polygonum*), especially *P. lapathifolium, P. pensylvanicum,* and *P. amphibium,* produce abundant seed crops, as well as providing important cover for wetlands and terrestrial wildlife (see Figure 4). Other *Polygonum* species either produce lower quantities of seed or production is irregular, but could be important in some regions.

Marsh submergents and floating types tend to have larger seeds and smaller tubers that are useable by many types of wildlife, especially waterfowl (see Figure 5). Notable are the various species of *Potamogeton,* some of which occur in most freshwater wetlands and almost all of which produce abundant seeds or tubers. Sago pondweed (*P. pectinatus*) is perhaps most well known, but most species produce abundant supplies of seed and tubers. Even the less obvious types may be important in some areas. In late summer and fall, the quiet back waters of New England rivers are often covered with uprooted leaves of the diminuitive *P. subterminalis* after puddle ducks have dug out and consumed the tubers. Succulent leaves and stems of *Myriophyllum, Ceratophyllum, Ruppia,* and *Elodea* support populations of more vegetarian species (gadwall, widgeon, swans, and coots) in many regions, whereas *Vallisneria* is used by muskrats, most ducks, and many geese wherever it occurs. Virtually all portions of *Nuphar* are sustenance to moose as long as the surface is ice-free and to muskrat and beaver throughout the year. Even the large seed of *Nelumbo* is used by some ducks, and it provides excellent brood habitat for young wood ducks in swamps, lakes, and reservoirs throughout the South.

Submergents also provide cover and feeding habitat for fish fry and dense beds in marshes, lakes, and rivers are often important spawning habitats. Protection from predators is important but the substrates for micro- and macroinvertebrates makes these beds critical to survival of larval and early immature stages of important freshwater fishes, as well as larval and adult stages of myriads of amphibians. Although emergents and floating leaved species also provide cover and substrate for invertebrate foods, dense stands of filiform-leaved submergents create abundant, optimal environments for invertebrates and immature fish and amphibians.

Floating species tend to be small and succulent, and entire plants along with vegetative propagules are food for birds, mammals, and fish. In addition, the microhabitats in floating mats of *Lemna* or *Spirodela* support abundant invertebrate populations that are heavily used by many vertebrates.

In wooded wetlands, submergents are limited or lacking, and trees and shrubs provide food and cover for wetlands as well as many terrestrial wildlife species. A few scattered, permanent ponds with dense populations of submergents, floating leaved, and floating herbaceous species may be inordinately important during the dry portion of the year to a variety of wildlife. The overall significance in terms of life support in

Figure 3.   This trumpeter swan brood will find relative safety from predators once it reaches the cover of the dense stand of emergents.

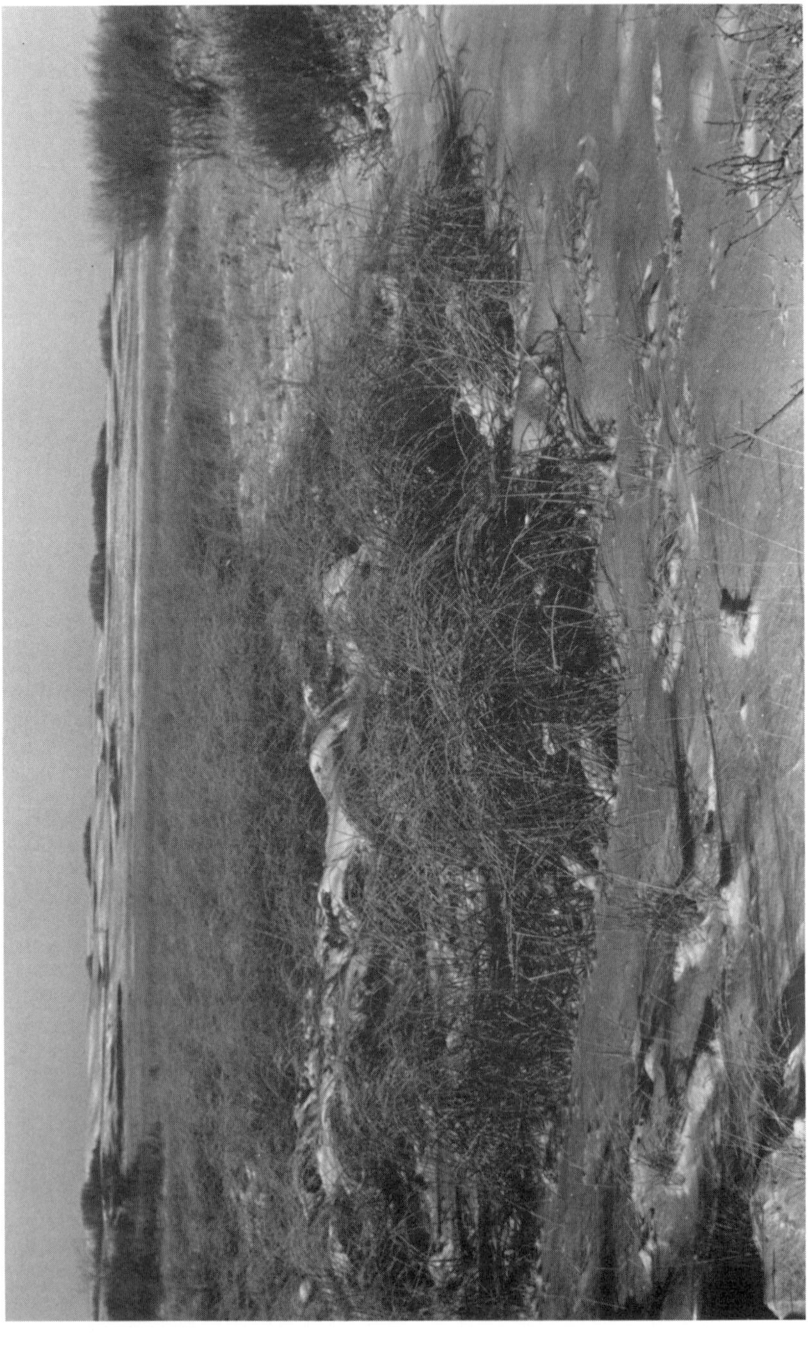

Figure 4.   Marsh vegetation provides important shelter for deer, fox, pheasants, grouse, and other upland species during harsh winter conditions.

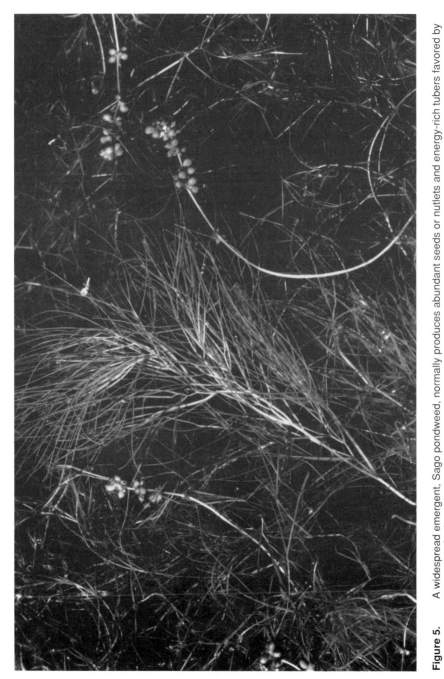

Figure 5.   A widespread emergent, Sago pondweed, normally produces abundant seeds or nutlets and energy-rich tubers favored by many waterfowl.

wooded wetlands is the hard mast and cover produced by shrubs and trees. A few species produce substantial quantities of soft mast or berries (blueberries, blackberries, cherries, hackberries, persimmon, grapes, etc.), but the food source that is significant to many important wildlife species is hard mast or nuts from oaks, hickories, and cypress. Though upland oaks produce large quantities of acorns in good years, depending upon the species, production may be limited to alternate or every 3 years. Many wet-growing oaks produce almost every year and seem much less susceptible to damaging spring frosts that frequently devastate acorn production of upland species. Doubtless, adaptation to life in low areas that often harbor pockets of cold air includes delayed flowering that avoids damage by late spring frosts. The result is that swamp oaks rarely fail to bear bumper crops that not only support wetlands wildlife, but frequently sustain terrestrial species, especially when upland oak production fails (see Figure 6).

Care is needed if fast-growing vines and brambles such as blackberry are used since brushy types tend to dominate and retard shrub and tree establishment and vines often smother young trees.

Small spring-borne seeds of buttonbush (*Cephalanthus occidentalis*) are food for many vertebrates, while its trunks and branches shelter fish, amphibians, reptiles, and birds, as do willows and birches.

Shrubs and trees not only provide shelter, perches, and open nest sites, but also create more cavity roosts and nests than their terrestrial cousins (see Figure 7). Fast-growing but short-lived birches break off, decay, and die, producing cavity nests for chickadees, nuthatches, and prothonatary warblers; sycamores provide comparable shelters for woodpeckers and wood ducks. Similarly, snags and windfallen oaks and hickories shelter deer, bear, and raccoon; and the base and lower branches of all harbor fish, amphibians, and reptiles, especially when water levels are high and flooding over the buttressed trunks of many aquatic trees or the knees of cypress.

Species selection for flood control wetlands should include a mix of marsh and swamp types. The initial planting would consist largely of marsh emergents - cattail, bulrush, rush - interspersed with fast growing willows, birches and ashes. If the area will be flooded to 2 ft or greater for more than 3 weeks during the growing season, *Pontederia* and *S. validus* will do better than *Typha, Juncus,* or the other *Scirpus* spp. As the new forest begins to develop, species representing later stages (oaks, hickories, gums, spruce, or tamarack) are put in place. Since the objective is to obstruct and slow flood waters, species with multiple stems or branches and dense growth forms are preferred. However, planners should carefully examine the anticipated flood depths, duration, and time of year in selecting candidate species from Tables 1 through 6 or other sources. Almost all species listed will endure extended inundation during the non-growing season, but many are intolerant of deep waters or flooding for more than 30 days during the growing season.

Only a few species have been widely used in wetlands designed for water purification: *Typha latifolia, T. angustifolia, Phragmites australis, Scirups validus, S. cyperinus,* and a few token attempts with *Iris, Pontederia, Acorus, Cladium jamaicense,* and various grasses (mostly *Phalaris arundinacea*). Cattail tolerates a wide range of water chemistries, including waters with very poor quality, and may be the

Figure 6.   Acorns of willow oak often carpet the forest floor, providing food for many kinds of birds and mammals.

Figure 7.   Old growth trees furnish shelter and perching and nest sites for wading birds, eagles, and osprey.

most commonly used in North America, though common reed has been planted in most European systems. Despite considerable discussion and limited results, a clear choice for the one best species is not apparent and may never be. Typically, only one or two emergent species are planted in most waste treatment wetlands, but many other species soon invade and it is not unusual to find 40 to 50 different species in a single system after only 3 to 4 years.

Water depth controls zonation of wetland plants; secondarily, water quality (especially turbidity and dissolved oxygen content) impact species distributions. Selection of species should be primarily based on planned operating depths, with chemistry considerations included if poor quality water is anticipated. Defining and mapping planting zones based on water depths in the new system facilitates species selection as well as determining quantity of planting materials needed. These zones are defined below.

1. Transitional: areas which are seasonally flooded but not inundated for extended periods during the growing season. These are commonly referred to as "wet meadows" or if wooded, "bottomland hardwoods." They are rarely flooded during the growing season, but water depths may be 0.5 to 1.0 m at other times of the year. As the name implies, this zone is a gradient between terrestrial well-drained soil conditions and saturated soil conditions in the next deeper zone.

2. Shallow: areas which are frequently or continuously flooded during the growing season with depths up to 10 cm. This zone occasionally dries during annual drought periods or may be dry for more than 1 year during extended droughts.

3. Mid: the zone that is continuously inundated with 10 to 50 cm. Generally, it only dries during prolonged droughts.

4. Deep: continuously flooded with 50 to 200 cm. Rarely dries.

Classifications of herbaceous wetlands plants in Table 1 and 2 represent the most frequently observed water conditions for robust growth of each species. However, some species of *Polygonum, Carex,* and *Phragmites australis* readily cross category boundaries, thriving in dry environments as well as in the transitional, shallow, and mid zones; many other species can survive in the near reaches of the neighboring zone. Latitude and altitude differences occasionally, and extremely wet or dry climates often, compensate for water level requirements, confusing attempts to categorize growing requirements of wetlands plants. However, under the appropriate water conditions, the majority of the plants listed in Tables 1 to 6 will survive throughout the mid-latitudes of the northern hemisphere and many occur from the treeline in the north to the equator in the south. In fact, some species have worldwide distributions (i.e., *Phragmites australis*) and many genera include very similar species present on every continent except Antarctica.

The broad adaptability and aggressive colonization ability of a few species may cause dominance of the entire system by only one of these weedy types. Unless the project is designed to provide treatment for wastewater or will receive poor quality water from some other source, developers should exercise caution in planting *Phragmites australis, Typha latifolia, T. angustifolia,* or any of the *Salix* spp. All are aggressive, weedy plants that easily dominate the entire wetlands if suitable growing conditions are present and all are difficult to control. In addition, in some regions, other species can easily become pests (e.g., *Paspalum*), as have many introduced wetlands plants. Obviously, exotic species should not be planted in the new system, regardless of project objectives.

Woody species shown in Tables 3 through 6 are grouped by tolerance to inundation during the growing season. Once again, latitudinal, altitudinal, and climatic differences blur category boundaries and project workers should remember that each category is defined as ability to tolerate flooding for a certain time period during the growing season. Most species will thrive under less severe flooding conditions and new plantings require more modest inundation for survival.

Other factors that need consideration include aesthetics, wind breaks, shoreline erosion protection, or other relative capabilities of one species versus another. In the end, plant species selection is reduced to reexamining the original objectives, eliminating those species that will not thrive because of climatic, hydrologic, or substrate conditions, and then selecting species that will provide the life support, flood buffering, or water purification functions that are required to achieve project objectives. As in site selection and design, compromises must be made without losing sight of the principal objectives of the new wetlands. However, water depth, and/or flooding duration, is the most important controlling factor for species survival, and selections should be based primarily on this criterion.

Since the objective is to create or restore a viable, self-maintaining wetlands ecosystem and the plant community is the basis for all other life in the system, species selection and planting provides an important opportunity to influence the diversity and complexity of the new system. Planners should attempt to create as diverse (in terms of number of species) and complex a system as feasible, since the self-maintenance atttribute of natural ecosystems is directly correlated with the degree of diversity and complexity.

As any farmer knows, maintaining a single-species system (a crop field) requires substantial inputs of time, energy, and chemical additives, and system failure is not unusual despite all his efforts. Simple systems lack resiliency that would prevent minor disturbances from causing major alterations to the system. For example, insect outbreaks in single-plant species system have devastated that one plant population with drastic effects on wetlands system performance. However, if the system has 20 to 30 species of plants, severe impacts to or the loss of one species will have much less impact on the functioning of the total system.

Consequently, planners should select and plant as many species within each zone as is feasible, recognizing of course that numerous species will naturally colonize and some of the planted species may be lost over time. Since most functional values of wetlands are dependent upon the interactions of many types of plants and animals with the abiotic components, establishing more rather than fewer types of plants in the initial stages will advance system development and subsequent performance of desired functions. At a minimum, 10 to 15 different species should be planted in the transition and shallow zones, and 5 to 10 species in the mid and deep zones. If natural sources for planting materials are used, core and/or soils associated with dug plants normally contains propagules of the plants occurring in the natural system and many different species will be planted. However, if commercial sources are used, a number of species should be obtained for each zone present in the new system instead of the common tendency to choose one or two familiar, showy or inexpensive species.

SOURCES OF PLANT MATERIALS

Planting materials may be purchased from nurseries, dug from the wild, or grown by the developers of a project. Advantages and disadvantages of these options influence quality of planting stock, general and seasonal availability of materials, and costs for acquisition and planting. Generally, plants collected from similar wetlands in the vicinity will be more suitable for most projects than plants obtained from other sources. Wild-dug materials will inevitably contain a considerable amount of seed and other propagules in the attached soil materials that will enhance establishment of a diverse, complex community of plants in the new system.

Plants collected from the wild are more closely adapted to local environmental conditions than those obtainable from any other source. Wild collections have acclimated to local soils, typical hydrologic regimes and, of course, regional weather patterns. Consequently, wild plants will initiate new growth more quickly and develop more robust growth habits at earlier stages than will plants secured from nurserys as seeds or potted plants, especially if the nursery is distant from the project. This is especially true in wetlands constructed for wastewater treatment or other harsh environments. For example, plants used in acid mine drainage treatment systems are generally collected from similar acidic environments because plants from normal habitats suffer considerable stress and need a lengthy acclimatization period before vigorous growth begins.

If collecting and planting are carefully scheduled, local collections will require limited storage capacity since the newly dug materials will be planted in the new system within a day or two. Digging is scheduled for 1 day, with overnight storage in damp, cool conditions followed by planting the following day. Depending on air temperatures and humidity, overnight storage may merely consist of a pickup truck with planting materials layered into the box, thoroughly moistened (but not so heavily watered as to wash soil from root hairs), and covered with burlap or synthetic tarps.

Digging plants is typically more expensive than obtaining them from some suppliers, but may be less costly than nursery-raised stock. Other disadvantages include availability concurrent with planting schedules because digging conditions may be difficult early or late in the growing season, undesirable species may be included in the seed bank, local supplies may be limited or difficult to obtain because of regulations or land ownership, and finally, careless collecting could impact natural wetlands or populations of rare or endangered species. Before digging, developers should secure a thorough biolgical survey of the collecting sites to insure that unique species or systems will not be damaged or that unwanted species will not be obtained. During collections, small plots (0.2 to 0.3 m) should be taken with a buffer zone of 0.5 to 0.7 m between collecting plots so that the undistrubed vegetation will colonize the collection plots within the same growing season.

Only a few commercial nurseries specialize in collecting or growing wetland plants.

Most of these are located in Florida, Maryland, New York, and Wisconsin. Other sources include commercial tree and shrub nurseries, seed and grass companies, state nurseries, and suppliers of native flower and landscaping plants. The district offices of the Soil Conservation Service (SCS) and university extension services have useful regional lists of plant dealers for conservation plantings, orchard and garden plants, and agricultural plants. Some of these include firms that offer plants suitable for wetlands projects. The Plant Materials Centers of the SCS occasionally have planting materials available for testing and demonstration, and are often familiar with sources for specific types or varieties.

Nursery-grown plants are supplied as potted materials, bare-root seedlings, container-erized seedlings, or bagged root balls with a short sapling in the case of many shrubs and trees. Potted materials are expensive, often $1.00 to 3.00 per plant, but are easily planted and usually suffer less transport and planting shock. For comparison, digging costs with union scale labor often fall in the $0.30 to $1.00 range, depending on substrate and equipment availability. Bare-root seedlings are typically $0.25 or less per seedling, but are much more susceptible to transport planting shock. Containerized seedlings may be enclosed in small plastic trays, in paper sacks, molded peat or wood fiber, and larger plastic containers. Avoid black containers if the summer is already warm and planting materials will be exposed to the sun. Cost varies with the type of container and size(age) of the plant, but most are several times more expensive than bare root seedlings. However, added costs may be justified because containerized seedlings will survive in sites that are to harsh for bare-root seedlings. Freshly collected plants from managed natural marshes available from a few sources vary from $0.15 to $1.00 per plant, and are not as susceptible to shock as bare-root seedlings. A few species of shrubs and trees may be obtained at no or low cost from state nurseries, depending on the state and the objectives of the project. Prices at commercial nurseries vary with the species and the size of planting stock, but typically range from $3.00 to $50.00 per seedling with most of the variation related to size.

Though costs may be higher, especially considering the need for express shipment, using commercially supplied plants has strong advantages in availability of large quantities and delivery on site at scheduled planting times. However, planners should contact potential suppliers well in advance (3 to 9 months) if they require large quantities (more than 500 plants) and to coordinate delivery dates and methods. Most suppliers can readily supply smaller quantities, but large orders must be started as seeds or collecting sites and methods determined well in advance of delivery dates. Large quantities should be shipped as several partial shipments to reduce on-site storage requirements and limit plant mortality. In addition, plants may or may not be available at the desired planting time depending on whether stock is grown outdoors or in greenhouses and local weather conditions. For example, planting conditions may be excellent in southern areas during March and April, but wild collections are difficult in northern latitudes until after ice-out, which may be as late as May. Greenhouse seedlings are usually available very early and late in the year, but generally are not available all through the winter, and seedlings from outdoor beds will often only be available during the first half of the growing season.

One significant disadvantage of using commercial suppliers is a tendency to plan wetland plantings based on the species available from those suppliers, and only a few species are commonly available. Intermixing wetland plants and soils from a regional site with the nursery-grown plantings will add diversity and complexity to the system without significantly increasing costs. Another disadvantage lies with the genetic and physiological adaptations of plants to their growing site that may inhibit their ability to survive and grow at other locations with different soil, hydrologic, and climatic characteristics. Developers should try to select a nearby nursery or, next best, a nursery with similar soils and climate conditions and should avoid large latitudinal distances between plant source and destination; longitudinal variation is more acceptable.

Planting materials from commercial sources should always be shipped by express package service, delivered by the supplier, or picked up by project staff. Although all of these will add cost, shipping delays of 1 to 2 days and/or storage under unfavorable conditions can cause substantial mortality with subsequent loss of labor and expenses for planting as well as replacement costs for planting materials.

Plants can be grown specifically for a particular project, but this often requires a year or more of lead time, and elaborate facilities such as a greenhouse or outdoor nursery. Few developers, unless they are planning a very large project or working on many projects over a number of years, can justify or sustain the additional effort and cost. In addition, propagation of wetland plants is currently much more of an art than a science with many failures to be expected during start-up of a new operation. Existing nurseries have overcome those hurdles and developed successful methods to propagate some species and can provide them at reasonable costs.

If costs, commitment, or interest favor establishing a facility to propagate wetland plants, information on seed scarification and seedling nutrient and water requirements for some species is available in the references on wetland plants in Appendix B. Additional sources include a number of scientific journals devoted to plants (botanical journals), the *Journal of the Society of Wetlands Scientists,* newsletters and journals of specialist groups for wastewater treatment with constructed wetlands and aquatic weed control, and a few regional journals covering specific areas with significant wetlands (i.e., the Chesapeake Bay or San Francisco Bay). After reviewing published information, developers should review the old annual reports from nearby U.S. Fish and Wildlife Service National Wildlife Refuges, and contact the Plant Materials Centers of the SCS and request advice and assistance from established nurseries, bearing in mind that the latter may be less than enthusiastic over helping a new competitor.

PLANTING

Since marshes are successional stages to bogs and swamps, efforts to establish either bogs or swamps should begin by establishing a marsh appropriate to the region. In addition, much more experience has been obtained in creating marshes than any other type of wetlands. Consequently, this discussion of vegetation establishment techniques concentrates on methods used in creating marshes. Planners developing a swamp on their site should first establish marsh vegetation and intersperse tree seedlings within the marsh plants, taking care to not use aggressive, tall, or heavy foliage, or emergent marsh species. Bog developers will want to distribute clumps of *Sphagnum* and other mosses and acidophilic herbs, as well as shrubs and trees, among the marsh plantings.

Most emergent wetland plants are perennials that reproduce and spread largely by vegetative means, their seed scarification requirements vary and, consequently, seed germination rates are quite low. However, many compensate for varying conditions and low rates by producing large numbers of seeds (e.g., *Typha, Phragmites, Scirpus,* and *Cyperus*). Conversely, many submergents reproduce prolifically by seed and a few have astonishingly long seed viability periods — in tens or even hundreds of years (e.g., *Potamogeton, Nuphar,* and *Nelumbo*). Others that may be rooted submergents or rooted with floating leaves or free floating produce vegetative structures, geminules, that survive harsh conditions on the bottom and establish new plants the following spring. Consequently, choice of planting techniques is strongly influenced by the type of wetlands and the plant species selected.

In general, the least expensive method, if a natural seed source is nearby or the substrate contains a seed bank, is to create suitable conditions for natural invasion and establishment. For most species, including many woody shrubs and trees, that consists of moist, almost soupy muds maintained by holding the water level at or immediately below the surface or by periodic shallow flooding and dewatering. Best results for typical emergents will be obtained by careful water level manipulation at the onset of warm weather in the spring.

Few emergents' seeds will germinate while completely submerged, preferring instead the warm, moist mudbars exposed after shallow waters recede leaving bare soil conditions. A few hardy terrestrial species will also establish during early phases, but as the wetlands species gain height, shallow flooding will inhibit or kill the terrestrial species, removing competition for wetlands emergents. At this point, deep flooding will also inhibit the growth and vigor of wetlands species and overtopping individual plants will likely cause mortality. Water levels must be carefully managed to maintain adequate depths to preclude terrestrial species, yet not impair the growth of the wetlands types. As growth continues and plants reach 0.3 to 0.7 m, water levels should be raised to optimal elevations for the desired species — usually 2 to 4 cm for *Typha, Scirpus, Phragmites, Cyperus, Sagittaria, Alisma, Pontederia,* and *Peltandra*.

Although the seeds of some submergents will germinate under shallow water, most require similar wet soil conditions for optimal germination and growth. Geminules of submergents and floating species are adapted to initiating growth while submerged, as are the large root stocks of the water lilies, *Nuphar, Nelumbo,* etc.

Assuming adequate seed sources, manipulating water levels to foster germination and growth of desired species and retard establishment of terrestrial types is substantially less expensive than planting, but may require 1 to 4 years to achieve full coverage. The time will depend on the type and size of a native seedbank, distance from a nearby seed source and seed dispersal mechanisms, soil fertility, and water management precision. In unusual cases, a dense stand of *Typha, Scirpus, Eleocharis,* or *Carex* may appear the first spring because a large quantity of seed was provided optimal germinating and growth conditions. More likely, the stands will be initially sparse, gradually increasing through the growing season and during the second year, but not reaching high densities until the third or fourth years.

Though less costly, using natural planting methods by manipulating water levels will generally take much longer to establish dense stands than would deliberate planting. In addition, planners may have limited influence or control over what species become established or their distribution within the new wetlands. To some extent, both types and distributions can be affected by water management related to elevation differences within each pool. However, these techniques are imprecise and results will vary. Conversely, the new system may be invaded by many species originating from seeds in the bottom soils or outside sources, even though most of the system is deliberately planted. Survival and spread of either planted or invading species will depend on which species are involved, soil fertility, water chemistry, and prevailing water depths.

Keep in mind that these species, as do most wetland plants, find their best growing conditions in moist soils but they have adaptations to grow in saturated or flooded soils where other plants are unable to survive. If the soils are kept too dry, terrestrial species will survive and compete with the emergents, reducing growth rates and inhibiting spreading by the wet types. The objective is to eliminate terrestrial species by flooding, but not cause stress on the wetlands species from deep flooding. Consequently, careful and precise water level management is important to successful establishment of vigorous, spreading stands of emergent wetlands species. Inability to flood uniformly and shallowly and dewater quickly because of bed slopes or inadequate water control structures will impede establishment of emergent plant communities. This is one important reason why bed slopes should be flat (i.e., 0.0% slope) and why fairly large controls with high flow capacity are needed.

Continual inundation into early or mid-summer, followed by rapid dewatering without periodic re-flooding will foster germination and growth of a completely different group of wetlands plants — the moist soil species.

Since wetland plants are infrequently exposed to drought conditions, most lack waxy surfaces, small leaves or thick coverings, and other water conservation measures that many terrestrial species possess. Consequently, disturbance and damage to fine root structures (root hairs) during transplanting frequently causes considerable stress because the roots are unable to take in enough water to offset the high rate of loss from the stems and leaves. Once new root development has occurred and water uptake ability is restored, wetland plants are able to thrive in moist soils; complete saturation or flooding is unnecessary. However, developers should anticipate seeing tops and, in some cases, portions of the root systems, die back in a high proportion of the new plantings. New growth will later begin from buds or root primordia on roots and rhizomes or shoots.

Deliberate planting times and methods vary with the species selected and type of planting materials; that is, seeds, tubers, rootstocks, or whole plants. In temperate regions, planting herbaceous vegetation is generally most successful in early spring, although the planting period extends from after the onset of dormancy in the fall to mid-summer. Though not recommended, cattail and bulrush have been planted later — in September in southern regions of the U.S. — as long as adequate time remains before killing frost to permit the new plantings to add new shoots. Root development is concurrent and continues after frost has killed the above-ground portions of the plants.

Tubers and root stocks are best planted in the fall after dormancy, but care must be observed with water levels and water quality. If the water quality is poor (i.e., low dissolved oxygen and/or high organic loading), submerged plantings may die from inadequate oxygen during the winter. Conversely, shallow water levels or simply wet substrates may freeze hard enough to kill tubers and root stocks if winter temperatures are extreme. Root stocks with 20 to 30 cm of stalk protruding above the water's surface allow higher water levels, even if the water is low quality, because the cut stalk provides a pathway for oxygen from the atmosphere to the roots. With good water, tubers of the most common emergent wetland plants, except cattail, are best planted after dormancy in the fall. Cattail and, to a lesser extent, *Phragmites,* and most of the grasses and sedges seem to develop and spread faster if planted immediately after dormancy is broken in the spring.

Broadcast seeding of trees and shrub seeds has had highly variable success, but hand-planting acorns and seeds of some other species is generally more successful. Pelletized seeds with coatings of fertilizer, fungicides, and stimulants are available for common species. Carefully handled bare-root seedlings, containerized seedlings, saplings, or cuttings are generally more reliable, though more expensive to purchase and more laborious to plant.

The planting window for woody species is considerably smaller. In most projects, fall and early winter plantings, after the onset of dormancy, have been the most successful. Some shrubs and trees can survive if planted in early summer, but the percentage lost is likely to be considerably higher. Since projects requiring wet-growing trees or shrubs are often in floodplains or other areas frequently flooded during the winter, choice of planting materials and water level management is crucial to seedling survival.

Flooding tolerance specifications are largely developed from observations on mature trees and shrubs, and seedling tolerance is frequently quite different. Not only do mature plants have growth form and adaptations (knees, buttresses, etc.) that protect them from flood damage, they also have height to avoid being overtopped by deep flooding. Seedlings overtopped for more than a few days during the first winter will probably not survive. In addition, moderate to high velocities (flow rates) are often associated with deep flooding that may extract or physically damage seedlings which lack a securing root structure. Consequently, if deep or prolonged flooding cannot be prevented, the extra investment in older, taller seedlings (possibly coupled with anchors) is good insurance.

Those that have observed a dense stand of cattail or other wetland plant suddenly spring up in a roadside ditch will probably question recommendations against seeding wetland plants. However, collecting seed and then distributing it over a suitable substrate has not been very successful, probably because some species have relatively low germination rates and most have stratification requirements that are not easily satisfied. Most species are capable of surviving extended periods of dormancy because of thick, hard, tight seed coatings that must be ruptured, dissolved, or decay before the embryo is freed. In some species, drying followed by immersion promotes germination; in others, drying reduces viability. Some require cold temperatures, others seem oblivious to temperature or storage conditions, surviving, for example, at room temperatures in dry museums for over 200 years! Others can survive after being frozen in ice or mud for much of the winter. Even within the same species and the same seed crop in natural habitats, germination tends to be erratic, with some seeds germinating after 1 year and others remaining dormant and viable for 4 to 5 years. Germination of the large tough seeds of *Nelumbo* spp. has been initiated by experimental treatment with acid to increase the permeability of the seed coat. Finally, as with most seeds, the proper conditions of moisture, temperature, oxygen, and often a light stimulus (perhaps UV radiation) are required for germination. With all the variables and contingencies, attempting to establish a new wetlands plant community by seeding is at best unreliable and at worst may simply fail.

Though seeding emergents and submergents is generally not successful, seeds of many wet-growing grasses have been collected from the wild or purchased from nursery sources and sown with good results. However, considerable variability has been noted in different regions. For example, reed canary grass (*Phalaris arundinacea*) is easily established by seeding in northern latitudes and, in fact, spreads prolifically in the Northwest, but is much less successful and most developers use plugs in the South.

Prior to sowing, the substrate is tilled or scarified, as for terrestrial plantings, and the seed is planted with a seed drill. If the seed is broadcast, the substrate should be tilled, raked, or dragged before and after seeding and before flooding. Most wetland grasses, exceptions include a few species of *Glyceria, Paspalum,* and *Panicum,* thrive on moist to saturated soils, without standing water, more typical of wet meadows than marshes. Consequently, the substrate should be shallowly flooded and dewatered repeatedly, or pool levels should be maintained at or immediately below the surface to foster germination and growth. The above exceptions grow in standing water or as floating mats and are generally better planted with root stocks or rhizomes.

Delayed seed germination is partially the basis for a fairly simple planting method that quickly generates a diverse plant community in the new system. Because of the large quantities produced and prolonged viability of wetland plant seeds, the substrates of most natural wetlands contain an abundance of seed of many species, commonly called the seedbank. Cores (10 to 12 cm diameter and 15 to 25 cm long) of wetland soil and included propagules (seeds, roots, tubers, and rhizomes) collected in existing marshes have been successfully transplanted to new wetlands in many regions. However, attempts to simply dig out wetland soils and spread the materials across the bottom of new systems have not been as successful. Consequently, core planting is laborious and costly due to the need for collecting, transporting intact, and installing the heavy cores. Also, the cores as well as surrounding substrate must be kept moist, as in any other planting method. But success with core plantings frequently develops a complex, diverse community in a very short time period. Additional advantages include: plants developed from propagules in good soil conditions do not suffer from shock and die-back commonly associated with whole plants, and the reservoir of seeds and other propagules provides a source for new plant development not only through the first growing season, but in some cases for years later.

Most emergents in new wetlands are started with hand planting of whole plants or dormant tubers and rhizomes. The root collar, evidenced by a dark line or a line dividing two color zones, provides a handy reference point for installing whole plants at the proper depth since it marks the previous soil line. It may seem obvious, but supervisors should emphasize that whole plants should be installed with the root down. Plants inserted upside down will develop slowly, if at all. Tubers are commonly forced into soft substrates deep enough to prevent them from floating out, while rhizomes are angled slightly upward in shallow slits or trenches and then covered over. In either case, the substrate should be well moistened, if not saturated, and must not be allowed to dry after planting.

Planting densities vary with target operating dates and species used throughout most projects use 0.75 to 1.5-m spacing for herbaceous vegetation and 3 to 6-m spacing for trees and shrubs. Closer spacing decreases the time required for spreading and complete coverage, whereas distant spacing requires longer periods for closure. Aggressive weedy species — cattail and giant reed — spread much more rapidly than rushes, sedges, bulrushes, arrowheads, and their relatives. Willows, alders, and other woody species that form colonies or clones spread rapidly through vegetative reproduction while oaks, hickories, gums, etc. that are dependent upon seed production and germination will not increase until initial plantings mature, perhaps 15 to 20 years.

Most often, developers use portions of or whole plants with shoots, roots, and rhizomes. If complete plants are used, the materials should be small, new growth transplanted in early spring. With later plantings when natural vegetation is taller, stalks should be cut to 20 to 25-cm lengths to prevent wind throw until the roots develop secure anchorage in the substrate. In either case, tops and roots of many of the grasses will die back and new growth is initiated from buds at the plant base. More hydrophytic species begin new growth from buds on the shoots and the pause is foreshortened. Transplanting shock is likely to be evident in both types, even though fairly large masses of roots and substrates are transplanted, and maintaining optimal water levels or soil moisture contents is critical.

Hand planting tubers, root stocks, or whole plants is labor intensive and costly. Although often recommended, weighting tubers and dropping them into the water is not overly successful. Placing tubers in weighted cotton mesh bags has been satisfactory in a few cases. Generally, a tree planting bar or tile spade is used to make a slit in the substrate, the propagule is inserted, and the slit is sealed. Plants should be placed so that the previous soil line (discoloration line on stalk) is level with the new soil line, but deep enough to prevent floating out when the area is flooded. Power augers are used to create holes for core plantings in dry soil, and spades are used if the planting is under water or in wet soil.

To reduce costs, a number of developers have used mechanical devices to expedite the planting process; simplest is perhaps using a trencher to cut a shallow ditch or row into which the planting material is hand placed and tamped by foot. Depending on substrate composition and stability, this method can substantially expedite planting. An additional advantage is that the bottom of the ditch is lower than the surface of the substrate and may be within the pool water level or at least will tend to collect and retain water longer than the top of the substrate, thus providing better moisture conditions for the new plantings. Farm equipment, modified to handle the species to be planted, such as tree planters, tobacco planters, and tomato planters also form a trench, place the root stalk or whole plant in the trench, and cover it. If the pool substrate will not be damaged by equipment and planners have access to any of these, planting times can be shortened and costs substantially reduced.

Depending on the principal function of the wetlands and planned water flow patterns, planting rows should generally traverse the narrow axis of the pond rather than the long axis. This is especially important for wastewater treatment systems because rows running the length of the cell will result in water flowing down each row, thus avoiding the retarding, filtering action of the planted vegetation. In some instances, flows with high volume and velocity have even prevented subsequent colonization of between-row areas.

Plant anchoring methods may be needed if the substrate is soft, planting materials are buoyant, or wave action or erosion will disturb the new plantings. Various erosion blankets (straw, coconut fiber, or synthetics) will protect substrates, hold plantings in place, and trap sediments to help stabilize the site. Barriers placed outward from the plantings to protect against wave erosion can be constructed of piled rock, drift fencing, or floating booms consisting of poles or logs anchored with chain or cable to the substrate. Netting pegged down at intervals will hold plantings in place and, depending on mesh size and placement, may provide protection against wildlife depredation. However, mesh may entrap wildlife, and synthetic materials are resistant to degradation and can easily injure wildlife. Mature cattail, bulrush, and reed plants receive little use by waterfowl, but tender new shoots with high protein content may be heavily grazed. Muskrats, nutria, and beaver feed on roots and tubers and may need to be excluded from new plantings with woven wire fencing.

Needless to say, plantings should be allowed to to become well established and overcome planting shock before other stresses are introduced. Depending on the objectives, expecting significant flood water retention, wildlife food production or wastewater treatment within the first growing season after planting is unrealistic and could cause plant mortality or system failure. If possible, stresses from flooding, wastewater, etc. should be gradually introduced into the system so the plants and other components have an opportunity to adapt to the new evironmental conditions. Adequate plant establishment could take 1 to 3 full growing seasons for herbacious vegetation and 10 to 20 years for woody species. During the startup period, the vegetation should be monitored frequently and if dead or unhealthy patches are found, they should be replanted and/or the introduced stress factor(s) reduced. Additional planting material may be obtained from healthy regions within the new system, from the original source, or from another source if poor growth or mortality is widespread. In the latter case, water level operations and other factors should be evaluated to identify causes prior to seeking another source for planting materials. Many wetlands plants are very adaptable and quickly become established in new environments if proper water and nutrient conditions are available. If substantial plant loss occurs, poor planting times or techniques, inappropriate water levels, or inadequate nutrients are more likely to be the cause than poor planting stock.

Planting tree seedlings should follow standard tree planting methods and seasons, but water levels will need to be high enough to ensure saturated soils but not so deep as to endanger the seedlings. Shrub and weedy species, *Salix, Cephalanthus,* and *Populus,* can be started with cuttings. Cuttings (0.5 m long and less than 2 cm diameter and including at least one vigorous bud) should be collected in the spring just prior to the end of dormancy. They may be stored in well-moistened sand overnight and then forced into saturated soils. Start-up management may require periodic flooding and drying during the first few growing seasons and shallow flooding during fall and winter. Remember, growth requirements for wet growing trees and shrubs are similar to herbacious wetlands plants. They have the ability to grow in much wetter environments than terrestrial species, but optimal growing conditions are moist to wet but not necessarily flooded soils. Conversely, maintaining dry soils will favor terrestrial species that may out-compete the planted wetlands types.

Herbaceous species of forested wetlands may be introduced by obtaining soil cores from an existing system and placing it at similar elevations in the new system. Frequently, the understory consists of species that exhibit maximum growth in early spring before canopy closure and many of these are available from wildflower nurseries.

Depending on adjacent wetlands or terrestrial habitats and animal populations, it may be necessary to protect shrub and tree seedlings from mouse, rabbit, or deer damage by placing wire screening around the new plantings.

Post-Planting Water Level Management

Proper water depth and its careful regulation are the most critical factors for plant survival during the first year after planting. Many plantings have failed because of mistaken concepts that wetland plants need or can survive in deep water. Small, new plants lack extensive root, stem, and leaf systems with aerenchyema channels to transport oxygen to the roots. Consequently, flooding often causes more problems for wetland plants during the first growing season than too little water, especially if the water has low dissolved oxygen content. In fact, optimal conditions for transitional species, emergents, and woody species consist of saturated, but not inundated soils. Submergents and floating species require actual flooding soon after germination or planting because most depend on buoyant structures and water pressure for physical support to achieve an upright growth form. The objective of water level management is to create unfavorable conditions for terrestrial species by shallow flooding or saturating the soil, but not to stress wetlands species by deep or prolonged inundation. Shallow flooding (1 to 2 cm) can limit invasion of weedy or terrestrial species once the wetlands plants have stems higher than 5 to 6 cm. However, it is absolutely essential that stems and leaves of desirable species project well above the water surface to avoid drowning new or even older established plants.

This is an important reason for designing systems with little or no slope on the substrate and easily maintained water control structures that precisely regulate elevations. Water level management must create very similar water depths and/or duration of flooding throughout the newly planted area, and must precisely maintain that level despite fluctuating inflows. Many new plantings have been lost because stormwater inflows could not be discharged rapidly enough to avoid overtopping and drowning small plants. Similarly, inability to drain the entire area on schedule due to undersized control structures or uneven substrates will cause spotty areas of poor growth or even mortality.

Current evidence suggests that gas transport from the atmosphere to root structures in saturated soils is influenced by a number of factors, including differences in the respective gas proportions (each individual gas moves towards the region with a lower concentration), similar gradients in relative humidity, and temperature differences between various parts of the same plant. In addition, some evidence suggests that wind blowing across the surface of leaves and stems creates a Venturi effect, pulling intra-plant gases out into the atmosphere. Replacement due to physical atmospheric pressure forces air into other sections of the plant and the combined effect establishes a flow pattern through the plant that is maintained until the wind dies.

Doubtless a combination of these factors operates in many species though one may be more important than another in certain environments. Regardless of which is paramount, access to free air above the water surface by portions of the plant is essential for any to operate, and managing that contact is the key to managing many wetland plants. Stems, leaves, "knees," and swollen trunks exposed to free air are analogous to breathing tubes; blocking those tubes will stress and eventually kill the plant. Since water level manipulation is the simplest method for exposing or blocking these breathing tubes, careful management is critical to survival of new plantings as well as older stands and, conversely, a method for limiting establishment and growth of undesirable species or restricting colonization into unplanned areas.

Wet meadow species that grow in the transitional or shallow zones should be watered during the first year by shallow (less than 1 cm) flooding with intermittent drying periods, depending on the species. Generally, transitional species can tolerate reasonably long dry periods, whereas shallow zone species should not be allowed to completely dry during the first growing season. Some species, such as various spikerushes (*Eleocharis*), many grasses (*Spartina, Festuca, Poa, Agrostis,* and *Alopecurus*), most ferns, and many herbs will tolerate extended dry periods. Others such as *Leersia, Glyceria, Cladium,* and *Carex* are not as tolerant and should not become completely dry the first growing season. A few, *Juncus, Polygonum, Cyperus, Paspalum, and Scirpus* can withstand 2 to 3 cm of inundation and complete drying of the surface as long as adequate soil moisture is present.

Water levels for woody species should be managed similar to those for transitional/shallow zone plants with the principle goal of shallow, intermittent flooding to restrict invasion or retard growth of terrestrial species but not impact wetlands types. Surface drying will not be detrimental if the soil has adequate moisture, but extended or deep flooding will retard the establishment of desired species. Bear in mind that the categories in Tables 1 to 6 are based on field observations of mature individuals of each species. Small, young shrubs or trees are not nearly as tolerant because they lack extensive root, stem, and leaf structures to supply adequate oxygen to their roots.

Emergent species (shallow to mid depth zones) should be planted in saturated but not flooded soils and allowed to grow stems with leaves that project above planned flooding levels the first season. After stems reach 5 to 10 cm, water levels can be raised 2 to 3 cm above the substrate and proportionately increased as plant height increases until desired elevations are reached. Unless otherwise dictated by project goals, best results will be obtained by maintaining 2 to 3 cm water depths throughout the new stand of emergents during the first growing season.

As soon as submergent and floating leaved plants show new and vigorous growth, water levels should be slowly and gradually increased to support erect, upright growth forms. Shallow overtopping is not detrimental, but exposure and drying should not be allowed to occur. Maintaining stable water levels and keeping the plant continuously submerged is critical for these species. However, submergents require oxygen, light, and nutrients similar to other types, but they must obtain oxygen from surrounding waters and receive adequate solar radiation to carry on photosynthesis even though completely submersed. Flooding the new plantings with turbid waters or waters with low dissolved oxygen will stress and perhaps cause mortality of submergents. Floating species or those with floating leaves can survive in poor waters so long as other conditions are suitable. Growth forms in many submergents vary with water quality conditions and, in poor water, they often have long, virtually bare stems with a profusion of leaves near the water surface. In good conditions, leaves are regularly distributed throughout the length of the stem unless water depths are too great.

Just as inappropriate water levels can inhibit establishment and growth of desirable wetland plants, unsuitable levels can be used to control prolific growth and spread of weedy species. Flooding may retard invasion by terrestrial opportunists and deeper flooding may retard undesired colonization of additional areas by planted species. For example, expansion of cattail stands may be controlled by maintaining depths of 30 to 50 cm especially if dissolved oxygen levels are low. Mowing or cutting the stems of many emergents, including *Typha* and *Phragmites,* followed by flooding well above the cut stems for several weeks during the growing season will thin or kill stands of these species. Basically, any method of interrupting the required oxygen or solar radiation supplies will stress and eventually kill most species.

If low temperatures and thick ice cover are anticipated, water levels should be slowly raised to accomodate ice thickness and reduce frost penetration into the substrate. In spring, for emergent species, water levels should be lowered and kept at or just below the surface of the substrate until new growth has reached 15 to 20 cm when levels may be raised to 2 to 5 cm above the substrate and gradually higher during the summer. For submergents, pool levels should be lowered in spring to 2 to 5 cm until new growth is evident, and then raised to normal operating levels.

SUMMARY

The principle goal of identifying candidate plant species and planting or fostering their establishment is to create as complex and diverse a plant community as practical in the minimum time interval. Complexity and diversity in the animal community is a direct function of a complex and diverse plant community, and the degree of diversity in the biological components governs the stability and self-maintaining capability of the total system. Deliberately establishing a diverse plant community will facilitate early system development and performance of wetlands functions, as well as reduce operating and maintenance needs.

11   ATTRACTING AND STOCKING WILDLIFE

We seem to have a natural tendency to think that since this is a new system and we have planted vegetation, we also need to introduce the kinds of wildlife and fish that we wish to have in the system. However, introduction should only be seriously considered after careful evaluation of factors affecting natural methods, the most important of which are type, distance, and connections with nearby natural wetlands.

Without a substantial commitment of time and effort, most introductions of native vertebrates will not succeed. Furthermore, most birds, mammals, fish, and reptiles, the showy species that everyone wishes to have immediately, are highly mobile, and if suitable habitats are present in the new system, these groups will naturally invade. If natural wetlands are nearby and birds, mammals, and turtles do not begin using the new wetlands, some essential requirement is lacking and no amount of introduction effort is likely to succeed. Fish will move in if the system is connected to other surface waters at almost any time of year, but especially in spring. Animals released in a new system with inadequate habitat will simply move to nearby wetlands that supply their needs or die.

Ecologists broadly refer to an animal's requirements with three terms that define important concepts: trophic level, niche, and habitat. An animal's habitat is its address, its home, its dwelling place, the physical location that provides cover or shelter, and opportunities to find food, water, and mates. Each animal is adapted to and requires a specific type of habitat and that is the address at which it can be located. Habitats might include grasslands, marshes, forests, deserts, and subdivisions of each. Its profession within that habitat, the manner in which it makes a living or how it obtains energy and nutrients (food) to survive and grow, is its niche.

Niches include one type of lichen growing on rocks, another type of lichen growing on tree bark, antelope grazing on forbs, deer browsing on twigs, bison grazing on grass, fox feeding on mice, coyotes feeding on rabbits, herons feeding on fish, raccoons feeding on carrion, fungi decomposing plant litter, and bacteria decomposing fecal material. Niches may be very narrow and specialized — snail kites (*Rostrhamus sociabilis*) living almost solely on apple snails (*Pomacea*) and night-blooming cacti pollinated by a single species of moth — or broad and generalized — snapping turtles (*Chelydra serpentina*) preying on invertebrates, fish, amphibians, birds, and mammals, foraging on aquatic plants and scavenging on dead plants and animals.

Broad categories of niches are grouped into trophic levels: plants are producers, herbivores are primary consumers, carnivores are secondary consumers, omnivores feed on plants (primary producers), herbivores (primary consumers), and perhaps carnivores (secondary consumers), decomposers (invertebrates, bacteria, and fungi) degrade organic material and, finally, others (parasites, scavengers, and saprophytes). Some organisms fit nicely into one trophic level; that is, most plants are producers since they create organic materials from inorganic substances, but even some plants are saprophytes, obtaining nourishment from dead plant material. Many microbes are decomposers, but some bacteria are chemosynthetic, capable of deriving their energy from chemical transformations and some may be decomposers or producers, depending on environmental conditions. Most animals blur the lines of distinction; that is, many animals (raccoons, coyotes, bears, and pigs) scavenge, feed on plants, and feed on other animals.

Producers, microscopic and macroscopic plants, provide the basis for most ecosystems, trapping and using solar energy to manufacture complex organic substances from simple inorganic materials. Though many plants would naturally invade a created wetlands, the last chapter addressed methods to facilitate establishing this important group to expedite operation of the system. Our purpose was to "jump-start" the new wetlands — to decrease the time period for establishing a diverse plant community. Natural methods of plant dispersal and invasion would have accomplished the same result, but they may have taken a bit longer. Since at some time of the year, most animal populations are larger than their habitats will support, individuals are forced to search for new homes and will discover the new system during their wanderings. Therefore, if suitable, unoccupied homes are available, most types of animals will discover the system within the first year of operation.

As plants begin producing potential foods for animals and transforming raw environments into suitable homes, animals from nearby systems will quickly discover the new habitats. Consequently, deliberate introductions are rarely necessary or warranted. Attempts to establish populations of animals are obviously doomed to failure until appropriate habitats and niches are available. If desired species occur in wetlands in the vicinity, but fail to appear in the new system, re-evaluate potential habitats and available foods, modify system parameters as needed, and allow time for pioneering before considering introductions.

In the worst case condition, the new system is remote (tens or hundreds of kilometers) from any similar natural wetlands or does not have any surface water connections with adjacent streams, lakes, or wetlands. Developers should then review the literature on similar systems, consult with local authorities, and select species for introduction that occur in natural wetlands, shallow lakes, or streams in the region.

Trapping or collecting animals, including insects and other invertebrates in some states, and many plants is prohibited or regulated by state or federal laws. Consult with the nearest office of the state natural resource agency on collecting locales, collecting methods, and permits required. Unless the project goal is to provide habitat for a unique (rare, threatened, or endangered) species, only commonly occurring, abundant animals should be collected from locations that will not jeopardize a unique species or population.

If planting materials were obtained from natural wetlands, most microscopic and many macroscopic organisms will have been included in soil cores and soil associated with plant roots. Assuming that soil cores were kept moist and cool, virtually all microscopic organisms and the eggs or immature stages of many invertebrates will have been introduced, along with a myriad of plant propagules, during the planting process. If not, obtaining and introducing soil cores is a practical means of bringing in these organisms.

Before thinking of introducing microorganisms, remember that they are ubiquitous, occurring and/or spreading rapidly into almost any environment that contains a source of energy and nutrients. In some cases, the environment may seem to be hostile; that is, acid mine drainage, but only the most extremely acidic waters (pH < 2) lack abundant microbial populations deriving energy by transforming metallic ions and nutrients from other inorganic sources. With the exception of a few very complex, large molecular weight compounds, some type of microorganism is capable of breaking down almost any substance known to occur in wet environments. A few naturally occurring and some anthropogenic polycyclic compounds are resistant but not immune to biodegradation, taking very long periods for decomposition and, in some cases, requiring initiation via radiation exposure. However, given time (remember, generation times may be only 20 to 40 h in some microbes), a population of microbes will develop that is adapted to degrading virtually any compound that contains a potential energy source. Consequently, regardless of the objectives for the new wetlands, inoculation of microorganisms is rarely warranted.

As with plant propagules, eggs and small life forms of invertebrates are often carried in the feathers, fur, or on the feet of birds and mammals. Even non-flying insects and mollusks may be injected into the system by a bird or mammal passing through. Many flying insects migrate over long distances and it would be very unusual if most did not appear almost overnight. Entomologists have long known that many northern insect populations are restored after hard winters by migrants carried north by spring winds. Occasionally transported at high altitudes, flying insects will quickly find and populate even the most remote system. A probable bumblebee (at least, it was yellow and black) smeared my windshield at 3353 m above sea level over Kentucky one spring afternoon. Many non-flying types hitch rides with migratory birds and wandering mammals.

If fish are desired, they most likely will need to be introduced. Mosquitofish (*Gambusia affinis*) are the most desirable because of their control on potential pest mosquito populations. Topminnows (*Pimephales*), shad (*Dorosoma*), shiners (*Notropis*), killifish (*Fundulus*), and other forage species are needed to provide food for game species. Pike (*Esox*) occur in many natural marshes and bogs, as do bullheads and catfish (*Ictalurus*), sunfish (*Lepomis*), bass (*Micropterus*), crappie (*Pomoxis*), and perch (*Perca*). Gar (*Lepisosteus*), bowfin (*Amia*), crappie, bullheads, and catfish are often found in southern swamps during periods of high water. Suckers (*Catostomus*), carp (*Cyperinus carpio*), or other bottom-feeders should not be used because their foraging uproots submergents and suspends fine sediments that block radiation for submerged plants. A few species may be obtained from hatcheries, but most will need to be trapped or netted, with appropriate permits from the state natural resources agency, in local water bodies and released in the new wetlands.

Amphibians (frogs, toads, and salamanders) and reptiles (turtles, snakes, lizards, and crocodiles) are archtypical of wetlands (see Figure 1). Amphibians lay their eggs in fresh water and reptiles lay bird-like eggs in moist earth or rotting vegetation. The simplest method of introducing each is to collect, with proper permits, egg masses of amphibians or egg clutches of reptiles in spring or early summer and place them in the new system. Most frogs and toads attach masses of eggs to stems and leaves or other underwater portions of living or dead plants. Simply cut off the stem, place it in a container with water from the wetlands, and later set the stem into similar water depths in a quiet backwater of the project. Most will survive a few hours in transit if the container is protected from warming. If longer periods are expected, large shallow containers with a high proportion of water surface for oxygen exchange should be used. Egg-laying times range from the first warm days of spring to midsummer, depending on species.

Many salamanders place their egg masses in moist layers or piles of organic debris that are difficult to locate without causing considerable disturbance to the natural area. Adults of many are aquatic and can be netted or trapped, similar to small turtles, but a portion of the trap must project into free air or trapped individuals will drown. Others can be hand-captured or caught with drift fencing and pits, transported in moist, cool containers, and released at similar sites in the created system.

Mudpuppies (*Necturus*) and a few relatives, commonly present in many natural wetlands, are unusual amphibians that attain sexual maturity and reproduce without transitioning into an adult air-breathing form. Immatures and adults can be netted or trapped similar to fish.

Freshwater and saltwater turtle eggs have been successfully dug up, transported, and placed in a simulated nest in well-drained but moist soil or sand with little or no vegetation. Incubation is dependent upon solar heating and a shaded area is unlikely to maintain warm enough temperatures (see Figure 2). Eggs may also be incubated in a pail of moist sand maintained at average soil temperature (6 to 8 cm below the surface) and hatchlings released in shallow waters. Commercial bird egg incubators are usually not satisfactory. Incubation periods vary from 30 to 150 days, depending on species and temperatures. Be careful in attempting to vary incubation periods with high or low temperatures unless you want only one sex; hatchling sex ratios of many species vary with incubation temperatures.

Nests of egg-laying snakes and lizards are typically found in moist accumulations of organic debris — rotting logs, brush piles, and clumps of herbaceous vegetation near or just below ground surface. Collect the eggs and duplicate the natural nest at the project site. Many snakes and lizards retain their eggs until hatching rather than depositing them in nests. Live-bearing snakes and lizards must be hand-captured or trapped (drift fencing with regularly spaced pits) and released in similar habitats.

Song birds, waterfowl, and perhaps wading birds are likely to be the first large animal visitors to the new marsh, depending on time of year. Migratory birds tend to return to the area where they learned to fly to nest and rear their young, and many show some degree of fidelity to their migratory stopovers and wintering grounds. If necessary, flightless young of many species and, in some cases, flightless adults can be used to stock new sites. Adult Canada geese (*Branta canadensis*) maintained in captivity for years and later transported to distant release sites for their first flight either

Figure 1.   Unless the new wetlands is far removed from any natural system, frogs, toads, and salamanders will quickly take up residence.

became resident in the release area or returned to it after spring migration. Ducks and geese for release can be purchased from private propagators or captured before they reach flight age in natural wetlands. In ducks and geese, females return to their natal area and males accompany their mate.

With a few exceptions, most waterfowl will discover and populate the new wetlands without assistance. Wood ducks (*Aix sponsa*), and probably other cavity nesting species, have such strong fidelity to natal areas that they often fail to pioneer into nearby suitable habitats. In fact, female wood duck ducklings banded in a nest box have been recaptured as nesting adults in that same box in succeeding years, and one female may use the same box for a number of years. Wood ducks can be transferred by capturing the female and hatching brood in a pole-mounted box, transporting box and contents to the new site, and placing the box on another pole before gently uncovering the opening. Some females will desert their broods and return to the trapping site, but others have a stronger attachment to the brood than to their original nest site. If the female deserts, developers will be faced with capturing and raising the brood before release. In either case, ducklings attaining flight at the new site will return to it for nesting (see Figure 3).

Figure 2.   Hatchling turtles emerging from underground nests follow open skylines leading them to the nearest water, and most new wetlands will be well supplied after a year or two.

Many private propagators raise a variety of ducks, geese, and swans, including many exotics. Developers should not be tempted to release exotic species in created or restored wetlands, regardless of how appealing. This cannot be over emphasized with regard to mute swans (*Cygnus olor*), the common swan that graces ponds and lakes in parks and zoos around the world. Though considered aesthetic by many, its aggressive territoriality often prevents any other species of waterfowl from using the same habitat, and it is not adverse to attacking pets and humans.

Many ducks are omnivores, feeding heavily on invertebrates in some seasons and using seeds or vegetation during others. A few, widgeons (*Mareca americana*) and gadwall (*Anas strepera*), are largely vegetarian on submergent plants and some diving ducks concentrate on invertebrates. Swans make extensive use of submergents, but also graze on tender young emergents and grasses. Geese tend to graze marsh or meadow vegetation, strip seed heads from grasses and sedges, and readily adapt to cereal grains, as do many dabbling ducks and swans. Canada geese thrive on pasture and golf course grasses so long as a farm pond or water trap is nearby.

Canada geese and mallards (*Anas platyrhynchos*) introduced or naturalized in an area closed to hunting will likely become pests. Both adapt to a variety of foods and living conditions and quickly lose their fear of man. Visitors exacerbate the situation by feeding them and soon their droppings litter neighborhood yards, golf courses, and swimming pools; neither should be introduced in urban environments, and casual visitors should not be encouraged to remain.

Wading bird (herons, egrets, ibises, and bitterns) populations have been restored to

Figure 3.   Extensive programs to erect and maintain nesting boxes have replaced tree
cavities lost to logging and restored wood ducks to much of their former range.

a few areas by raising and releasing young or, in one instance, by holding adults in large,
seminatural cages for a number of years. Most are colonial nesters and are unlikely to
take up residence in a new system unless existing colonies are nearby, even though
migrants and dispersing young may visit the site for many years (see Figure 4). The
same principle, releasing young at or slightly before flight age, appears applicable to
some and should be tested on other species. Cormorants (*Phalacrocorax*) have been
attracted to new sites by placing nesting platforms in snag trees in one area, and
flightless young brown pelicans (*Pelecanus occidentalis*) were used to restore an island
nesting population at another. In each case, restoration efforts have entailed substantial
commitments of time and resources.

Wetlands raptors, osprey (*Pandion haliaetus*), and bald eagles (*Haliaeetus leu-*

Figure 4.   Great blue herons and other wading birds will locate a new wetlands but are unlikely to establish nesting colonies until shrubs or trees have matured to provide suitable nesting substrates.

cocephalus) have also been restored by raising and releasing young birds, but these projects also are expensive, time consuming, and must be continued for many years.

If necessary, wetland mammals can be live-trapped and released in similar habitats in the created system. Large numbers of inexperienced young make trapping and restocking easier in late summer and early fall than in other seasons. In addition, many mustelids — mink (*Mustela vision*), weasel (*Mustela* spp.), and otter (*Lutra canadensis*) — have delayed implantation, and females caught in the fall are likely to be pregnant and will give birth at the new site next spring. Even skunks (*Mephitis* or *Spilogale*) may be live-trapped, the trap gently covered with burlap, and the trap slowly and smoothly lifted for transport in a pickup bed to the release site. Mice, voles, and shrews are readily captured in small live traps, but shrews must have abundant food available during captivity to supply their high metabolic needs or they soon starve (see Figure 5).

This brief outline of introduction methods should not be interpreted as encouraging deliberate efforts to stock animals in created or restored wetlands. It is included merely to demonstrate that under worst-case conditions, many wetlands animals can be stocked using existing techniques described in the references or in recent journals. Most animals that can survive in the new system will naturally colonize it; if they do not invade naturally, introductions are likely to fail because suitable habitat is not available.

Microorganisms will be introduced along with transplanted vegetation and soil and by birds and mammals, as will many larger invertebrates. If the system has surface water connection with existing streams or other water bodies even if only during high flows in the spring, many fish and aquatic insects will rapidly colonize it. Most reptiles and amphibians will gradually move in from nearby systems; turtles reside in farm ponds many kilometers from the nearest water body. Birds will suddenly show up some morning; many migrate in late evening hours or at night. Wandering young mammals dispersing from home sites will follow stream courses or travel overland to an isolated marsh. Only with very unusual circumstances will introductions of any type of animal be warranted and, then, only after thorough determination that suitable habitats exist, but seed populations are remote or isolated by an impassable barrier. Fortunately, stocking is rarely necessary, or else creating or restoring a wetlands for life support functions would be a very expensive, long-term undertaking.

Figure 5.   Mice and other small mammals follow bands of dense riparian vegetation, leading them to new habitats. Many are important foods for larger mammals and birds, and in addition further diversity and interest for recreational users.

Natural ecosystems, including wetlands, are largely self-maintaining; that is, the complex of animals, plants, soil, water, and air that make up the total system perpetuates itself through time with only minor if any changes occurring as the result of disturbances or perturbations. The old-growth forest continues as an old-growth forest for hundreds and even thousands of years as do prairies, deserts, tundras, and many other natural systems. Over time, the animal and plant components have adapted to and modified the abiotic components (parent soil materials, hydrology, climate, atmosphere, and radiation) slowly establishing a complex of interacting components that is resistant to change. If a major disturbance impacts the system (e.g., a hot fire), thereby eliminating major structural components, the remnants begin the process anew and, through a series of successional stages, restore a very similar if not identical system.

This process of plants and animals adapting to and modifying the abiotic components and, in so doing, establishing increasingly complex systems with the more complex stage gradually replacing its less complex predecessor, is known as *ecological succession*. The final, self-perpetuating combination of living and non-living components is referred to as the *climax stage* or condition, and each interim state is known as a *successional stage*. Increasing complexity within each stage is generated by increasing diversity; that is, greater numbers of different species and fewer individual members of each species, resulting in rapidly increasing numbers of connections or pathways between species. A direct result of increasing interactions through more and different connections is the stability of the overall system or the inherent ability to resist modification or change despite impacts from outside factors.

Most types of wetlands are not climax stages but, in fact, represent successional states (see Figure 1). Largely because of our relatively short life span, we casually observe a cattail marsh over 50 to 60 years and tend to think of it as a stable, self-perpetuating system — the marsh is still a marsh. However, depending on the region, closer examination would reveal subtle yet irreversible changes on-going in the marsh all the while. In the absence of major impacts or disturbance, species of plants and animals and numbers of individuals in each species change from year to year, and the overall pattern of change is predetermined and predictable. Basically, the depression gradually fills, water depth decreases, and plant and animal communities change from those adapted to deep water through those of shallow water and eventually to those that

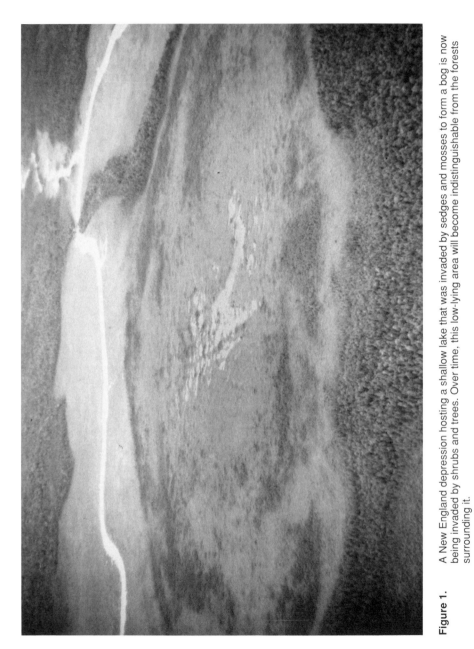

Figure 1.   A New England depression hosting a shallow lake that was invaded by sedges and mosses to form a bog is now being invaded by shrubs and trees. Over time, this low-lying area will become indistinguishable from the forests surrounding it.

thrive in dry, terrestrial environments. In the pothole region of the Dakotas and adjacent provinces, the terrestrial plants are grasses and forbs and animals adapted to living on these plants — bison, elk, antelope, prairie dogs, ground squirrels, and mice. The muskrats, ducks, fish, and wading birds that lived directly or indirectly on marsh plants are gone. The marsh has become a prairie. The new stage is, in this case, the climax under existing climatic conditions and the prairie will perpetuate itself until major climate changes occur. In more humid climates, the marsh may gradually become a bog, perhaps a swamp, and eventually a forest. The wetlands systems of interest to us are merely an ephemeral, transitional state if viewed in the total context of ecological succession.

Viewed from the perspective of geological time, the overall process of gradually succeeding stages culminating in the climax form in either case is predictable and irreversible; but examined from a short-term human viewpoint, the marsh persists for many years and appears to be a self-perpetuating stable system. Without an understanding of the changes that have gone on before and the changes that will inevitably occur in the future, we perceive the marsh (or any other wetlands system) as a permanent feature of the landscape, an enduring combination of living and non-living elements interacting to perpetuate the whole. Managing a new wetlands system with the misconception that it is an unchanging, stable, or permanent system will at least cause needless concern for the managers and, at worst, lead to inappropriate management changes and excessive costs. Even within the short human time frame, considerable change will and should occur. The astute manager understands the inevitability of change and the need for appropriate disturbance and interrupts, and reverses or influences the manner of change to accomplish project goals and the functional benefits sought from the new wetlands (see Figure 2).

To this point we have caused major changes in the hydrology of our site to create conditions suitable for wetland plants and animals, and we have introduced selected members of those communities. The new wetlands are presently in a very early stage of succession and now our management objective is to foster those changes that will increase the diversity and complexity of the wetlands, yet prevent or reverse changes that will cause replacement of the wetlands by a more advanced successional stage. We hope to shape the new system so that it develops the structure and function of a mature, stable, self-maintaining system, within our perspective, that we visualized in formulating our earliest plans. Doubtless, those plans were rooted in observations of a nearby wetlands or perhaps engendered by impressions from literature, films, or talks with others; but the model system had adapted and changed through many years until reaching the state that we now view. Certainly, we should not expect the newly created or restored system to immediately duplicate the structure and function of model wetlands. Unfortunately, many do. Many are disappointed that the new wetlands are not identical to model wetlands in the first year or two! Many projects have "failed" because the site was graded, flooded, and planted, and then left to its own resources, and the diversity and complexity of a well-developed system was still absent 2 or 3 years later.

Figure 2.   Turtle tracks highlight the drought impacting this wetlands. Although many wildlife species will be temporarily displaced, most will profit from the rejuvenated productivity following reflooding.

In the first place, it is grossly unrealistic to expect to create even the simplest type of natural wetlands systems in that short time; secondly, deserting a new wetlands and expecting it to suddenly become a fully developed system reveals a limited understanding of the basic factors that create and maintain natural wetlands. If the hydrology is correct, many "failures" may well become example wetlands, but development could require tens or hundreds of years; 2 or 3 years is a bit short, but had the developers actively manipulated the single most important factor (hydrology) they could have substantially increased system development and made good progress towards creating a complex, diverse, and relatively stable system. Active management of the new system is needed to expedite progressive development to the desired successional stage and then arrest further development through simple but critical manipulations of driving factors, principally hydrology.

Directional management in the form of water level manipulations is crucial during the first growing season, irrespective of the type of wetlands. After planting is completed in the spring, the emergent vegetation area of the pool should be inundated with 2 to 3 cm of water for 5 to 7 days to inhibit germination and/or retard growth of opportunistic terrestrial species. This level will saturate but not inundate the soils in much of the transition zone, and depths in deeper areas (the submergent/floating leaved zone) will range from 5 to 150 cm, depending on topography. Water levels are then lowered to at or just below the surface of the emergent zone substrate and maintained at that elevation for 15 to 20 days. By this time, evidence of renewed growth of planted materials should be apparent throughout the transitional, shallow, mid, and deep zones in the system; emergents can be expected to have produced shoot lengths of 5 to 10 cm. If soils from other wetlands were introduced, seeds of many additional species will germinate, creating a green carpet throughout the zone. The water level is raised to flood the emergent zone slightly deeper, (3 to 5 cm for another 5 to 7 days) and then lowered for 15 to 20 days as before. If significant mortality of planted material has occurred, leaving large unvegetated regions, levels should be lowered to the substrate elevation or just below it and new plants installed as needed during this period (see Figure 3).

By midsummer, most emergents should have stems in excess of 30 cm, seed bank species should be above 15 cm, and submergents/floating leaved species proliferating across the deeper zone. However, the transitional zone likely contains a substantial number of and perhaps high proportion of terrestrial species that are beginning to outcompete the desired wetlands plants. In addition, the emergent zone may now contain a significant proportion of transitional or moist site species that became established during the second drawdown period. Depending on stem height of wetland plants that came in with soil materials, water levels should be raised to 10 cm or higher (15 cm if it will not overtop desired plants) above the emergent's substrate for 3 to 5 days, increased to 15 cm for 3 to 5 days, and eventually to 25 cm. During this gradual raising, care must be taken to insure that the water surface does not overtop desired species. The objective is to gradually raise the level, following increases in stem height of desirable species without overtopping the latter.

If possible, gradually increase the depth to between 30 and 50 cm in the emergent zone and maintain that level for 10 to 20 days. Regardless of the final elevation,

Figure 3.   Wetlands plants are capable of surviving relatively dry conditions without undue stress, although a prolonged dry spell would allow invasion of terrestrial species.

carefully monitor plants in the shallow to mid zones (basically, the emergents) for signs of stress, indicated by reduced growth rate or in severe cases, yellowing or chlorosis. If stress becomes apparent, immediately reduce water levels to 2 to 3 cm over the emergent substrates. Any additional plantings that were added during the drawdown will be more susceptible to stress and should be carefully watched during these flooding periods. The goal in raising levels is to inundate as much of the transition zone as possible, flood and kill terrestrial species in the transition and emergent zones, yet not overstress and retard growth in the emergent zone. Submerged and floating leaved species also need to be carefully monitored, though most will have little difficulty with deeper waters unless high turbidity or low dissolved oxygen are potential factors.

After 20 days, reduce levels to 5 to 7 cm over the emergent substrates and maintain that elevation until late summer. Water levels should be gradually increased to 10 to 15 cm over emergent substrates, well before the onset of cold weather and higher if thick ice cover is expected. During winter, water depths should be managed to maintain free water below the ice cover or at least avoid freezing the substrate and damaging roots and rhizomes. However, the water or ice surface should not overtop stems of emergent species or oxygen supplies will be cut off and substantial mortality may occur. Do not dewater the system at anytime during freezing temperatures.

For wooded wetlands, water level manipulations should follow a similar pattern, though depths and duration of flooding must be less. During late spring and early summer, flood the area to depths of 2 to 3 cm for 2 to 3 days, followed by 10 to 15 days of drawdown, and repeat the cycle. In late summer, flood the new wetlands to 3 to 5 cm for 5 to 6 days and then dewater for the remainder of the summer. After leaf fall, gradually raise water levels to 2 to 3 cm for 5 days, dewater for 10 days, followed by 3 to 5 cm for 5 days, dewater for 10 days, and then maintain 10-cm depths for 5 days, alternating with 20 day drawdowns for the remainder of the winter. Be careful of lengthy drawdowns during freezing temperatures or the shallow root system of newly planted shrubs and trees may be damaged. Again, the goal is to create unfavorable conditions for terrestrial species without overstressing wetland plants by using the principle that flooding conditions are not optimal for wetlands trees or shrubs but they can tolerate inundation better than terrestrial types. Depending on the region of the country, normal winter rains may well create the winter flooding/drying conditions that favor wetlands species and simplify or confound management efforts (see Figure 4).

During the first warm spells in spring of the second year, raise the water level to flood most of the transition zone and maintain that level until warm weather is firmly in place and new growth started (may require close examination) on last year's plantings; then lower water levels to at or immediately below the surface of emergent substrates for 5 to 10 days, raise levels to 1 to 2 cm over the substrate for 2 to 3 days, and lower them again to the surface for 5 to 10 days. The goal is to create warm, moist mud flats to enhance germination and growth of any wetlands species brought in with soil materials during planting. As new growth of these plants becomes evident, slowly raise water levels to 1 to 2 cm over the substrate and then gradually higher until depths in emergent zones are 8 to 10 cm. Do not overtop stems of any desirable new growth during this period. With one exception, maintain 8 to 10-cm depths for the rest of the summer. In late summer when emergents have adequate stem height and growth is

Figure 4.   Most forested wetlands thrive on winter flooding but will only tolerate short periods of inundation during the growing season.

slowing, increase water levels in the pool to flood at least the lower half of the transition zone to depths of 2 to 4 cm and maintain that level for 5 to 7 days, and then return to normal operating elevations.

Fall and winter operation is similar to the previous year. During the third summer, manipulate water levels as in the second year. In the fourth spring, flood the transition zone early, lower water levels to 2 to 4 cm above emergent substrates for 10 to 20 days, and then raise levels to normal operating levels (8 to 10 cm) for the remainder of the summer. Fall and winter operation is the same.

Procedures for the fourth year become normal operations for subsequent years if the project is destined to establish a marsh. Sequentially raising and lowering levels during the first few years fosters invasion and/or growth of additional plants introduced with soils or carried in by wind, birds, or other seed dispersal agents. If natural wetlands soils were incorporated or if natural wetland are nearby, the new system should have 40 to 60 species of wetlands plants by the end of the third growing season. Do not be alarmed if one species proliferates rapidly one year and then fails to appear the next. This commonly happens with the smaller *Eleocharis* spp., many floating types, and of course, rooted or non-rooted algae. Wild rice (*Zizania aquatica*) is notoriously fickle, but in fact seems to vary with those all-important warm, moist, mud flat germinating conditions in early spring. Algae blooms wax and wane with dissolved nutrient conditions. If they become a problem, dewater during summer to flush the system, maintain moist but not flooded soils, and reflood in the fall. After deep flooding in early spring, dewater as quickly as possible to flush out redissolved nutrients and limit future nutrient inputs.

More commonly, weedy species (*Typha, Phragmites, Salix,* or *Populus*) will proliferate and soon dominate the system, crowding out other species and reducing system diversity. If extensive areas need control, cut or mow stems of *Typha* and *Phragmites* at the soil surface during a drawdown in late summer and then flood as deeply as practical over the winter. Complete control may require treatment over 2 or more years. As an alternative, consider blasting potholes in dense emergent stands with the ammonium nitrate-fuel oil mixture used by waterfowl managers for years. This inexpensive mixture is now commercially available, but it is a high explosive and should only be used by competent personnel in safe areas (i.e., not in an environmental education wetlands surrounded by housing developments or commercial businesses!) Stems or trunks of *Salix, Populus,* or *Cephalanthus* need to be cut and the stumps painted with an herbicide labeled for near or in water use (Rodeo); again, more than one application may be necessary.

Purple loosestrife (*Lythrum salicaria*) is a native of Europe that was introduced during early settlement and has spread into most wetland habitats north of the 35th parallel. It is a highly aggressive weedy species able to colonize and dominate the transitional and shallow to mid zones in virtually any marsh and even open-crowned wooded wetlands. Moderate control is possible with flooding to overtop the stems for 5 weeks or longer during the growing season, but complete control will likely require multiple applications of a broad-leaved herbicide (glyphosate or Rodeo). Drawdown and mudflat exposure during the first half of the growing season in an infested area will produce an explosive growth of purple loosestrife. Seed collections from some commercial sources contained purple loosestrife only a few years ago; so if commercial seed is used, specifications to exclude purple loosestrife need to be included with the order and shipments should be carefully screened before planting.

In more southerly regions, water hyacinth (*Eichhornia crassipes*), Brazilian elodea (*Elodea densa*), Eurasian watermilfoil (*Myriophyllum spicatum*), hydrilla (*Hydrilla verticillata*), Australian pine (*Casuarina equisetifolia*), tamarisk (*Tamarix* spp.), and other introduced exotics have become serious pests in natural wetlands. Complete control of most is only possible with repeated, proper application of recommended herbicides by weed control specialists.

Depending on regional soil fertility and nutrients carried in by inflows, simply manipulating water levels at the appropriate times may sustain a complex, diverse and productive marsh for many years. However, in most circumstances, available nutrients become bound in reduced conditions in the substrates and basic productivity declines. First indications are typically the proliferation of water crowfoot (*Ranunculus* spp.) and relatively clear water, followed by the appearance of bladderwort (*Utricularia* spp.) as nutrient supplies become limiting. The marsh should be dewatered and dried out as much as possible during one growing season. If purple loosestrife is a concern, delay the drawdown until late summer. If not, and project goals are life support, recreation, or education, do not despair over the lost season. Dewatering in early summer will cause an explosive growth of transitional species (*Polygonum, Bidens, Eleocharis,* and *Echinochloa*) in the dewatered regions that produce large quantities of seed used by many types of wildlife. Invertebrate populations also respond to nutrients from decomposing vegetation and those released by oxidizing the substrates, and the combination will attract spectacular populations of wildlife when the system is reflooded in the fall. Nutrients released from plant and animal decomposition and from substrate "sinks" are available for use by all segments of the system the following spring, supporting dramatic rebounds in productivity.

Created or restored bogs and fens must be managed quite differently, avoiding extreme hydrologic disturbances that benefit marshes. In either a bog or fen, the determining factor appears to be the pH of the water; water levels should be similarly fluctuated during the first and second growing season, except depths must be much less. Mosses are just as readily stressed or killed by prolonged flooding during the growing season, as are emergent plants; since most mosses have heights of less than 10 cm, flooding depths should be correspondingly lower. In addition, prolonged flooding during the growing season will be detrimental to introduced shrub and tree species. Finally, low water levels must not expose moss root structures or desiccation will cause stress and mortality.

However, similarly manipulating water levels during the first 2 years will foster establishment of a variety of species, some of which will later die out but some will remain, enhancing the overall diversity of the system. After the second winter, water levels should be lowered to and maintained at normal operating elevations for the duration of the project. In contrast to marshes, bogs require low nutrient, acidic but stable, constant-elevation waters. Fluctuating water levels, increased nutrient contents, and/or decreased hydrogen ion concentrations will favor marsh species, shifting the bog towards a marsh. Opposite conditions in these parameters reverses the trend and a marsh will gradually take on the characteristics of a bog. With adequate water control structures, maintaining stable water levels is not difficult, but limiting nutrient inputs from runoff may require pretreatment in a small marsh upstream from the bog.

Fens occur in similar habitats as, and occasionally surround bogs. They are outwardly more similar to marshes, but waters in most fens are very low in nutrients and relatively stable. Many fens appear to be located in groundwater discharge areas; hence, the low nutrient and hydrogen ion concentrations. They share other characteristics with marshes in that dominant vegetation is an emergent form (typically *Carex* spp.) and pH of the water is neutral to slightly alkaline. Low (2 to 5 cm), stable water levels with low nutrient content and moderate alkalinity will enhance growing conditions for fen vegetation. Limiting nutrient input from runoff is an important management principle and adding lime to decrease acidity may be required, depending upon pH of influent waters and precipitation. Since both could become difficult to implement over a long period, attempts to establish fens at locations other than groundwater discharge areas should be cautiously undertaken.

Long-term management of wooded wetlands consists of simulating normal winter flooding and summer drying with an occasional flooding (less than 5 days) during the active growing season; typically, bud eruption to mid summer, but slightly later for some shrubs. Water level fluctuations during the first few years will increase diversity by permitting transitional and some terrestrial species to become established in minor high spots, microhabitats, within the swamp. If terrestrial types proliferate, increase the frequency but not the duration or depth of flooding during the growing season until after desirable species are approaching maturity. Conversely, if wetlands species do poorly, investigate soil moisture conditions during dry periods and/or decrease flooding frequency, duration, and depths during the growing season (see Figure 5). Unless goals are to foster the most flood-tolerant species, complete inundation in the winter should not exceed 15 days, with drying periods of equal or greater length. If only highly tolerant species are desired, increase flood frequency and duration during the growing season after 8 to 10 years of growth and carefully monitor vegetation for signs of stress. In most cases, even the moderately tolerant species can withstand considerable flooding for one season, though stress will be evident (leaf color, leaf fall, and bark cracking). But two or more flooded growing seasons will cause substantial mortality.

Shallow wetlands in regions with prolonged hot, dry periods may lose considerable water with gradually receding levels falling from shallow shorelines. Summer storms that restore levels may appear to bring a reprieve, but managers must be cautious, due to the ever-present danger of botulism under these conditions. Warm, moist mudflats with high organic contents and perhaps other poorly understood factors seem to create suitable environmental conditions for bacteria-producing botulinum toxins. Shallow reflooding of previously exposed mudflats apparently dissolves and disseminates these extremely toxic compounds, affecting a wide variety of vertebrates. To avoid catastrophic losses, wetlands managers may need to either flood potentially dangerous mudflats with deep water, immediately reduce levels below pre-rainfall levels, or implement scaring/hazing methods to prevent wildlife use.

Controlled burning can be a useful tool to open up dense stands of vegetation or shape composition of plant communities; for example, to foster grasses at the expense of forbs, reduce accumulations of organic matter, and restore hydraulic capacity and

Figure 5.   Excessive depths and duration of flooding during the growing season caused substantial mortality in this well-established swamp, allowing buttonbush, river birch, and other weedy species to invade much of the area.

influence insect and mammal populations. In older systems, hot fires can reverse successional changes by reducing peat deposits and resetting the system back to its earliest state.

Obviously, these recommendations are perfectly applicable to the ideal system in which each zone — transition, shallow, mid, and deep — is flat-bottomed and the only variations in substrate elevations are between zones. Few natural wetlands exhibit this pattern and few if any projects will have the natural site relief or the funding to create these conditions. Water level changes will not and need not be instantaneous because of limits on size of water control structures and maximum inflow/discharge rates. However, this lag results in slow increases and decreases in water depth, and coupled with the ranges for flooding depths and duration in the above, creates extended, repeated periods with warm, moist but not inundated soil conditions that are optimal for germination and growth of wetlands plants. Once these plants are firmly established, water levels are gradually raised and maintained at normal operating elevations for the rest of the summer to inhibit invasion of terrestrial species.

MOSQUITOS

Mosquitos inhabit almost all wetlands, but population size and inherent problems vary substantially with type of wetlands and region of the country. Largely because of previous human experience, a few mosquitos may cause serious problems for wetlands managers in the South and West, whereas clouds of biting pests are accepted as normal living conditions in the Midwest and Northeast. The latter have extensive natural wetlands producing high populations of mosquitos, and human populations have learned to accept the inevitable. After dark, outdoor activities are simply not possible; but for example, in Tennessee, my neighbors will complain to the mayor that evening, not next morning, if bitten by a single mosquito while in the yard after dark!

Common misconceptions aside, the largest populations and most aggressive types seem to occur in cold, high latitudes and not in the tropics. I have experienced only minor mosquito problems in the Carribean and in the Brazilian rain forest towards the end of the rainy season. However, encountering unending clouds of mosquitos on the first warm, calm day of the summer in Churchill, Manitoba brought a quick understanding of shorebirds perching on any elevated structure and gulls nesting high in trees!

Potential mosquito problems vary strongly with distance to human residences and activities since foraging range varies significantly between different species of mosquito. In addition, females tend to avoid shaded waters for egg-laying, prefer calm rather than moving waters; typical egg to adult cycles range from 5 to 9 days and highest populations occur in organically rich waters. Control methods include deep flooding to strand flotsam and debris in the spring, repeated dewatering to strand mosquito larvae before they metamorphose into adult stages (i.e., on 5-day intervals), system design and vegetation management to preclude stagnant backwaters with no or limited connections to the main pool, shading the water surface, introduction of mosquitofish (*Gambusia affinis*) and insuring they have access to the entire water column, careful monitoring and, if needed, removal or dispersal of floating mats of *Lemna, Spirodella,* or other floating species and introduction of bacterial control (*Bacillus thuringiensis, B. sphaericus*). If all else fails, proper application of appropriately labeled insecticides may be necessary, but caution is advised. Overzealous insecticide application has stressed emergent vegetation in at least one system.

Suitable system designs, minor vegetation or water level management, and introduction of *Gambusia* are inexpensive preventive measures that should be incorporated in any system where minor mosquito production will cause adverse reactions by society. However, before initiating more involved procedures, carefully inspect or enlist the assistance of a good vector control specialist to inspect every aspect of the wetlands system. This includes such minor components as an old soda pop can, discarded tires, undrainable depressions in wooded areas, hollow stumps, water control structures, open piping, and anywhere else that you can possibly imagine that standing water might occur. Quite likely, problem mosquito production originates from some small but frequently overlooked component of the system or auxiliary artifact. Identify and correct these sources prior to initiating an extensive control program and, if the latter is chosen, enlist the services of specialists.

SYSTEM PERTURBATIONS

Although many wetlands, especially marshes, thrive on disturbance or change and astute management will deliberately introduce perturbations as needed, additional variations in "normal" operating conditions will occur more often than predicted or preferred. These perturbations can be categorized as: (1) periodically occurring and predictable and (2) infrequently or rarely occurring and hence unpredictable but probable. Predictable disturbances can be anticipated and were hopefully incorporated in earliest planning. If not, modifications may be possible during system startup. In contrast, unpredictable disturbances confront managers with serious and potentially catastrophic situations. In a sense, all disturbances that were not included in system planning and design can be considered unpredictable since good planning has considered all potential problems and accommodated those that are practical. However, designing for all possible events, for example a 1000-year storm event, would needlessly increase costs, but one could theoretically occur at any time.

Predictable disturbances include seasonal changes in precipitation, runoff, nutrient and sediment inputs, groundwater levels, stream or river flows, evapotranspiration rates, air and water temperatures, vegetation growth patterns and animal activities, and in some areas inflow or discharge changes because of upstream use, increased discharge to sustain aquatic life in receiving waters or comply with water rights regulations. Most should have been included in the designs, but one or more was likely overlooked and will need to be accommodated during start-up. In contrast, unpredictable perturbations include record storm events, vegetation damage, failure of a dike, water control or other structural component, significant changes in adjacent land use or water rights, design flaws, and pest or disease outbreaks.

Predictable disturbances are likely to be first encountered during start-up. In addition, system start-up periods provide opportunities to test and debug mechanical, electrical, and hydraulic systems and assure adequate performance. Soil amendments or chemicals to modify water chemistry (such as lime or fertilizers) may be needed to stimulate initial growth of macrophyte and microbial populations. High precipitation events test capacity of water flow control devices and water recording instruments, but may cause channeling that should be corrected with vegetative or mechanical baffles or weirs. Conversely, droughts test water retention capacity (including the liner) and endurance of plant and animal communities. Many predictable disturbances are seasonal perturbations that are specific to local climate, soils, waters, and vegetation and operating procedures can generally be modified during start-up to incorporate those that were overlooked during design.

Unpredictable disturbances may be serious problems since only limited capacity has been included to accommodate them. Record storms, earthquakes, or tornados, sudden appearance of a sinkhole in the pool or any other event that damages physical structure such that the entire system is jeopardized are all improbable but could create the need for extensive and costly repairs. A gradual increase in residential housing may not be predictable in the design stage, but should be identified as a potential factor influencing future operations. Construction activities might also cause episodic and excessive inputs of sediments that may smother the system. Heavy foraging in early spring during a one-time use by a large population of migrating geese could seriously retard annual growth, and if it occurs in the earliest years, may necessitate replanting. If the same flock used the wetlands for loafing between feeding forays to nearby fields, the nutrient additions could initiate or accentuate algal blooms later that summer. Unusual insect outbreaks have devastated wetlands vegetation and were only brought under control with insecticides. Wild fires could destroy significant proportions of the living biotic component and decomposing materials, causing substantial changes in the hydrology and biology of the system. Extreme and/or prolonged drought may cause operational changes or even require additions from other sources such as groundwater. However, do not be tempted to use irrigation return water regardless of the severity of the drought. Deposition of irrigation water contaminants in wetlands soils could have long-term, far-reaching detrimental impacts.

If the project objectives are to establish a flood buffering or wastewater treatment wetlands, initial introductions of flood water or wastewater could cause considerable disturbance. Planners should gradually phase in low to moderate levels of flood waters and dilute concentrations of wastewaters during the first years of operation lest the abrupt introduction create a serious perturbation that hinders full system development or detrimentally impacts new plant communities.

This discussion of disturbances is not meant to be an all inclusive list of minor deviations from normal operating conditions or probable catastrophes. It is included to stimulate planners to identify as many potential disturbances as possible, categorize each as predictable or unpredictable, and to choose whether or not to incorporate provisions for each in the system design and operating procedures. Most predictable events are recurrent and need to be accommodated, whereas the rare, improbable, and unpredictable events will need to be dealt with as they occur.

At this stage in the technology, every created wetlands is an experiment, with the exception of classic waterfowl marshes; but as each new disturbance arises and is dealt with and as each new idea is applied and the results monitored and recorded, the accumulated information base will support derivation of predictive relationships and wetlands creation will slowly become more of a science than an art. Consequently, a long-term monitoring program and detailed records on operations, disturbances, results, and modifications is the responsibility of every wetlands developer and manager.

ROUTINE MAINTENANCE

Since system hydrology is the most important and most easily manipulated factor, maintaining control and monitoring system inflows, outflows, and water levels is essential to managing the new wetlands. Insuring integrity of dikes, berms, spillways, and water control structures should be a regularly scheduled activity. Each of these components should be inspected at least weekly and immediately following any unusual storm event. Any damage, erosion, or blockage should be corrected as soon as possible to prevent catastrophic failure and expensive repairs.

Vegetative cover on dikes and spillways should be maintained by mowing and fertilizing as needed. Frequent mowing encourages grasses to develop good ground cover and extensive root systems that resist erosion and prevents establishment of shrubs and trees. Roots of the latter could create channels, with subsequent leakage or even dike failure. Muskrats and other burrowing animals can damage dikes and spillways and unrepaired burrows may lead to dike failure. If wire screening was not installed in the dikes, a thick layer of gravel or rock over trouble spots may inhibit burrowing. However, if damage continues, trapping and shooting may be needed for temporary relief until wire screen can be installed. Burrows are most easily repaired by setting an explosive charge from the top of the dike to collapse the network of tunnels and then filling the subsequent crater with compacted clay. Fences, pathways, roadways, and visitor facilities should be inspected concurrent with the weekly dike inspections and repaired as necessary. Pesticides or other chemicals should not be used unless extreme circumstances warrant use, in which case care is necessary to avoid damaging the system. This also applies to insecticides, since heavy applications of insecticides have damaged emergent vegetation in some wetlands.

Livestock grazing may cause serious damage to wetland vegetation, especially during the early years when plants are becoming established. Rubbing and loafing beneath shrubs and trees often damages new growth and accelerates soil erosion, exposing roots and causing mortality. Perimeter fencing may be required if livestock are anticipated to be a problem.

MONITORING

Wetlands are complex, highly productive systems with diverse and abundant populations of animals and plants interrelated in myriad pathways between biotic and abiotic components. To attempt to measure and understand every component and each pathway for energy and nutrient flow is far beyond the scope of normal operating guidelines. Yet because our knowledge of the components and processes is incomplete, we lack universal, readily identified indicators of system robustness and viability or, conversely, danger signals. Early diagnoses of failing functions are difficult to recognize, the most appropriate adjustments are not well understood, and the results of alterations may not be evident for long time periods. Consequently a long-term monitoring plan is essential to develop an information base for continuous comparisons of functional status and biological integrity of the system.

The plan need not be elaborate or lengthy, but it must provide clear documentation of monitoring objectives, organizational and technical responsibilities, specific tasks, methods and basic instructions, quality assurance procedures, schedules, reports, resource requirements, and costs. Since the life of the project may easily span many decades and numerous personnel changes, written documentation is essential to insure that data sets are at least comparable if collection or analysis procedures change, as is likely to happen. A carefully defined monitoring plan should be part of the operations manual so that it is readily available to serve as a benchmark for data collection throughout the life of the project.

Recorded measurements of water surface elevations, and inflows and outflows in each unit of the wetlands are basic monitoring parameters to correlate and understand changes in survival and distribution of biological components, especially macrophytes. Water levels are generally measured with staff gauges located in the deepest part of a pool within easy viewing distance of a dike or on a sidewall of a water control structure. Gauges should be positioned so that readings represent elevations above mean sea level and can be directly related to substrate elevations so that areal extent and volumes can be calculated from the "as built" construction drawings map of the system. Inflow and discharge can be similarly measured with staff gauges in simple "box" or V-notched weirs at each location. If considerable variation is anticipated and knowledge of flows is critical to maintaining desired water levels, automated flow measuring and recording devices (hydrographs, etc.) may be needed. Under normal steady-state operating conditions, weekly readings of pool levels, inflows, and discharges are adequate.

During start-up in all systems and during operation of wooded systems, determining and recording flooding frequency, duration, and extent of coverage is important. This is basically the frequency and time period that the water surface elevation is above the elevation of the wetlands substrate in any part of the system. Surface water elevations, determined from a staff gauge or recording graph, are related to substrate elevations on the "as built" topographic map to calculate flooding depth, frequency, and duration of flooding in each area and the volume of the water in the wetland. The latter is useful in predicting the time needed to flood the system to a certain elevation with a given rate of inflow or to dewater it with available discharge rates.

Understanding change in the system largely means understanding fluctuations in the biological communities as a result of hydrologic changes or some other perturbation. Monitoring viability and robustness (i.e., the health and well-being of the system and its individual components) requires a comparative baseline for that system or comparative information from a similar system. Disturbances may originate internally as well as externally and may be biological; that is, invasion of a new species, as well as physical or chemical (i.e., temperature changes or variation in quality of inflows). In addition, changes in one parameter may be manifested by single or multiple, and indirect as well as direct cause and effect alteration in an apparently unrelated component or process. Because of the tremendous number of interactions, the web of pathways for energy and nutrient exchange between species, between species aggregates or communities comprising a trophic or functional unit, and between biotic and abiotic components, monitoring basic parameters such as energy flux or nutrient exchange is impractically complex and expensive. Consequently, monitoring programs attempt to identify and collect information on potential indicators that are expected to rapidly reflect changes and are relatively easily measured. Selected macrophytes, vertebrates, and macroinvertebrates are common components of most wetland monitoring plans.

Wetland macrophytes, rooted, and floating plants are the primary producers supporting all other life in the system. Consequently, knowledge of changes in the macrophyte component are essential to understanding changes in any other component or process, as in Figure 6. Wetland vegetation is subject to seasonal, annual, and long-term changes in species composition and in distribution; that is, location and extent of coverage within the system. Some species may change little from year to year, while others may flourish for a year or two and then disappear. Monitoring these changes requires information on which species are present at what locations in the system during a specific time interval. A number of methods are used to determine each, but most are basically enumerations of individuals of each species along line transacts randomly or regularly bisecting the system or from quadrants distributed throughout the system. In either case, the sample points should be permanently fixed locations sampled with the same procedures at the same times of year. In addition, color aerial photos and color aerial IR photographs are used in conjunction with ground sampling to map species distribution and changes from year to year. IR photos detect different chlorophyll contents of different species, as well is in diseased or infested individuals, providing early indications of stress in many plant species. Measurements of standing crop biomass from clipping or harvest sampling and samples of litter/duff material and organic content of the substrate from core samples are useful in understanding decomposition, mineralization, and accumulation rates.

The large vertebrates are probably the most commonly sampled components next to vegetation, primarily because of their prominence and our interest in their populations. Unfortunately, very few vertebrates make good indicators and most types are unlikely to reflect minor but perhaps significant long-term changes in the system. Birds are sampled with direct counting methods that attempt total counts of the entire wetlands or selected portions (see Figure 7). Mammals, reptiles, and amphibians are indirectly sampled through track counts, scent posts, or trap, mark, and recapture methods. Fish are collected with gill nets, trap nets, or electro-shocking devices. In contrast to other vertebrates, fish (especially bottom feeders) are exposed to contaminants in the sediments and analysis may provide early warning of accumulation and bioconcentration.

Besides plants, invertebrates are perhaps the best indicators of health and well-being in a wetland system. Macroinvertebrates are sensitive to environmental stress, relatively long lived, and resident in the system occupying nearly all levels in the trophic structure. Taxa diversity and abundance respond rapidly to environmental stresses and sampling methods are simple and easily conducted. Commonly used methods range from grab samples to artificial substrates for colonization over a period of time and later

Figure 6.   Monitoring plants is relatively simple yet provides easily interpreted indication of basic changes in the wetlands system.

Figure 7.   Birds occupy a variety of niches and are easily monitored indicators of system
            health and well being.

collection. Analysis is dependent upon accurate identification since changes are
detected by changes in types and numbers of different taxonomic groups. Some groups
are readily identified, but others may require considerable training or the assistance of
a specialist. Many macroinvertebrates also serve as early indicators of contaminant
problems since many are detrital feeders or scavengers.

   Managing a wetlands system to maintain a healthy and functional community of
aquatic plants and animals requires feedback on changes that occur in components and
processes. Monitoring selected communities over time provides a comparative basis
for judging condition and response to disturbances or management changes. A well
thought out and rigorously implemented monitoring plan is an essential component of
the operating guidelines for any wetlands system.

EQUIPMENT

Annual Buyer's Guide
Water Environment & Technology
Water Pollution Control Federation
601 Wythe Street
Alexandria, Virginia 22314
703/684-2400

Ben Meadows Company
3589 Broad Street
P.O. Box 80549
Atlanta (Chamblee), Georgia 30366
800/241-6401
404/455-0907

The Birds' Nest
7 Pattern Road
Bedford, New Hampshire 03102
603/623-6541

Forestry Suppliers, Inc.
Box 8397
205 West Rankin Street
Jackson, Mississippi 39204
601/354-3565

Gundle Lining Systems, Inc.
19103 Gundle Road
Houston, Texas 77073
713/443-8564

Isco, Inc.
P.O. Box 82531
Lincoln, Nebraska 68501
404/474-2233

VEGETATION: PLANTING STOCK AND SEEDS

Hastings
434 Marietta Street, N.W.
P.O Box 4274
Atlanta, Georgia 30302
404/524-8861

Seeds of Change
621 Old Santa Fe Trail, #10
Santa Fe, New Mexico 87501
505/983-8956

Stock Seed Farms, Inc.
R.R. #1, Box 112
Murdock, Nebraska 68407
402/867-3771

Burr Oak Nursery
Route 1, Box 310
Round Lake, Illinois 60073
312/546-4700

Gardens of the Blue Ridge
P.O. Box 10
Pineola, North Carolina 28662
704/733-2417

Environmental Seed Producers, Inc.
P.O. Box 5904
El Monte, California 91734
213/442-3330

Beersheba Wildflower Gardens
Beersheba Springs, Tennessee
615/692-3575

Vick's Wildgardens, Inc.
Box 115
Gladwyne, Pennsylvania 19035
215/525-6773

Sharp Bros. Seed Co.
Healy, Kansas 67850
316/398-2231

Savage Farms – Nurseries
P.O. Box 125
McMinnville, Tennessee 37110

Wildlife Nurseries
P.O. Box 2724
Oshkosh, Wisconsin 54901
414/231-3780

Environmental Concern, Inc.
P.O. Box P.
St. Michaels, Maryland 21663

Van Ness Water Gardens
2460 Euclid Avenue
Upland, California 91786

Slocum Water Gardens
1101 Cypress Gardens Road
Winter Haven, Florida 33880

Mangrove Systems, Inc.
504 S. Brevard Avenue
Tampa, Florida 33606

Lilypons Water Gardens
Lilypons, Maryland 21717

Hawkersmith and Sons Nursery, Inc.
Route 4 — Box 4155
Tullahoma, Tennessee 37388
615/455-5436

Kester's Wild Game Food Nurseries, Inc.
P.O. Box 516
Omro, Wisconsin 54963
414/685-2929

APPENDIX B   REFERENCES

Adriano, D. C., and I. L. Brisbin, Jr., Eds. *Environmental Chemistry and Cycling Processes* (U.S. Department of Energy, Savannah River Ecology Laboratory; NTIS, CONF-760529, 1978).

Alexander, M. *Introduction to Soil Microbiology* (New York: John Wiley & Sons, Inc., 1967).

Allen, H. H., and C. V. Klimas. "Reservoir Shoreline Revegetation Guidelines," Technical Report E-86-13, U.S. Army Engineer Waterways Experiment Station (1986).

"Ann. Ponds — Planning, Design, Construction," USDA, SCS, Agriculture Handbook No. 590 (1982).

"Ann. Biological Field and Laboratory Methods for Measuring the Quality of Surface Waters and Effluents," Office of Research and Development, U.S. EPA Report-670/4-73-001 (1973).

"Ann. Wetland Habitat Development with Dredged Material: Engineering and Plant Propagation," U.S. Army Engineer Waterways Experiment Station, Tech. Report DS-78-16 (1978).

"Ann. Water Measurement Manual, Water Resource Technical Publication," No. 2403-00086, U.S. Government Printing Office (1974).

"Ann. Wetlands: Their Use and Regulation," Office of Technology Assessment OTA-0-206, U.S. Government Printing Office (1984).

"Ann. Impact of Water Level Changes on Woody Riparian and Wetland Communities," Vol. I-VI, USDI, Fish and Wildlife Service (1977-78).

"Ann. Final Environmental Statement — Operation of the National Wildlife Refuge System," USDI, Fish and Wildlife Service (1976).

"Ann. U.S. Army Corps of Engineers Wildlife Resources Management Manual," Environmental Lab., USA, Waterways Experiment Station).

Beal, E. O. *A Manual of Marsh and Aquatic Vascular Plants of North Carolina* (Raleigh, NC: North Carolina Agric. Exp. Sta. Tech. Bull. No. 247, 1977).

Bishop, E. L., and M. D. Hollis, Eds. *"Malaria Control on Impounded Water,"* U.S. Government Printing Office (1947).

Coleman, J. M. "Dynamic Changes and Processes in the Mississippi River Delta," *Geol. Soc. of Am. Bull.* (July 1988), pp. 999-1015.

Cox, G. W. *Laboratory Manual of General Ecology* (Dubuque, IA: W. C. Brown Publishers, Inc., 1970).

Cowardin, L. M., V. Carter, F. C. Golet and E. T. LaRoe. "Classification of Wetlands and Deepwater Habitats of the United States," USDI, Fish and Wildlife Service, FWS/OBS-79/31 (1979).

Cronquist, A. Holmgren, Revell. *Intermountain Flora — Vascular Flora of the Intermountain West, U.S.A.* Columbia University Press.

Cross, Diana H. (compiler) *Waterfowl Management Handbook,* USDI, Fish and Wildlife Service, Fish and Wildlife Leaflet 13 (1988).

Dennis, W. M., and B. G. Isom, Eds. "Ecological Assessments of Macrophyton Collection, Use, and Meaning of Data," (Philadelphia: ASTM Publication 8453, 1983).

Ewel, K. C., and H. T. Odum, Eds. *Cypress Swamps* (Gainesville, FL: University Presses of Florida, 1984).

Farb, P. *Face of North America* (New York: Harper & Row Publishers, Inc., 1963).

Flint, R. F. *Glacial and Quaternary Geology* (New York: John Wiley & Sons, Inc., 1971).

Gale, W. F. "Botton Fauna of a Segment of Pool 19, Mississippi River, Near Fort Madison, Iowa, 1967-1968," Iowa *State J. Res.* 49:353-372 (1975).

Garbisch, E. W., Jr. "Highways and Wetlands, Compensating Wetland Losses," USDOT, FHA; FHWA-IP-86-22 (1986).

Good, R. E., D. F. Whigham and R. L. Simpson. *Freshwater Wetlands* (New York: Academic Press, 1978).

Greeson, P. E., J. R. Clark and J. E. Clark, Eds. *Wetland Functions and Values: The State of Our Understanding* (Minneapolis, MN: American Water Resources Association, 1978).

Hammer, D. A., Ed. *Constructed Wetlands for Wastewater Treatment* (Chelsea, MI: Lewis Publishing, 1989).

Haslam, S. M. *River Plants* (Cambridge: Cambridge University Press, 1978).

Hook, D. D., and R. Lea, Eds. "The Forested Wetlands of the Southern United States," USFS, SE Forest Experiment Sta. Gen. Tech. Rep. SE-50 (1989).

Horwitz, E. L. "Our Nation's Wetlands: An Interagency Task Force Report," U.S. Government Printing Office, 041-011-00045-9 (1978).

Hutchinson, G. E. *A Treatise on Limnology,* Vol. I (New York: John Wiley & Sons, Inc., 1957).

Hutchinson, G. E. *Limnological Botany* (New York: Academic Press, Inc., 1975), p. 660.

Kadlec. J. A., and W. A. Wentz. "State-of-the-Art Survey and Evaluation of Marsh Plant Establishment Techniques: Induced and Natural, Vol. I: Report of Research", Vicksburg, MS, Technical Report DS-74-9, U.S. Army Engineer Waterways Experiment Sattion (1974).

Kusler, J. A. "Our National Wetland Heritage: A Protection Guidebook," (Washington, DC: Environmental Law Institute, 1983).

Kusler, J. A., and G. Brooks, Eds. "Wetlands Hydrology," (Berne, NY: Association of State Wetland Managers Tech. Rep. 6, 1988).

Kusler, J. A., and M. E. Kentula, Eds. *Wetland Creation and Restoration: The Status of the Science: Vol. I and Vol. II,* (Washington, DC: U.S. EPA, EPA 600/3-89/038a, 1989).

Kusler, J. A., M. L. Quammen and G. Brooks, Eds. *Proceedings of the National Wetland Symposium: Mitigation of Impacts and Losses* (Berne, NY: Association of State Wetland Managers, 1988).

Leedy, D. L., R. M. Maestro and T. M. Franklin. "Planning for Wildlife in Cities and Suburbs," USDI, Fish and Wildlife Service (1978).

Leedy, D. L., and L. W. Adams. "A Guide to Urban Wildlife Mangement," USDI, Fish and Wildlife Service (1984).

Lewis, R. A. (Creating Saltwater Wetlands). (Ann Arbor: Lewis Publ., 1991).

Linduska, J. P., Ed. *Waterfowl Tomorrow,* (Washington: USDI, Fish and Wildlife Service, 1964).

Majumdar, S. D., R. P. Brooks, F. J. Brenner, and R. W. Tiner, Jr. *Wetlands Ecology and Conservation: Emphasis in Pennsylvania* (Easton, PA: Penn. Acad. of Sci. Publ., 1989).

Mason, H. L. *A Flora of the Marshes of California* (Berkeley: University of California Press, 1969).

Mitsch, W. J., and J. G. Gosselink. *Wetlands* (New York: Van Nostrand Reinhold Company, 1986).

Nelson, R. W., G. C. Horak and J. E. Olson. *Western Reservoir and Stream Habitat Improvements Handbook* (Ft. Collins, CO: USDI, Fish and Wildlife Service FWS/OBS-78/56, 1978).

Odum, E. P. *Fundamentals of Ecology* (Philadelphia: W. B. Saunders Company, 1971).

Poston, H. J. *Wildlife Habitat: A Handbook for Canada's Prairies and Parklands* (Edmonton: Canadian Wildlife Service, 1981).

Reed P. B., Jr. *National List of Plant Species That Occur in Wetlands: National Summary* (Washington: USDI, Fish and Wildlife Service Biol. Rep. 88(24), 1988).

Riemer, D. N. *Introduction to Freshwater Vegetation* (Westport, CT: AVI Publishing Co., Inc., 1984), p. 207.

Sanderson, G. C., Ed. *Management of Migratory Shore and Upland Game Birds in North America* (Washington: Int'l. Assoc. of Fish and Wildlife Agencies, 1977).

Sculthorpe, C. D. *The Biology of Aquatic Vascular Plants* (London: Edward Arnold Ltd., 1967).

Sharitz, R. R., and J. W. Gibbons, Eds. *Freshwater Wetlands and Wildlife* (USDOE, Savannah River Ecology Laboratory; NTIS; CONF-8603101, 1989).

Smith, H. K. "An Introduction to Habitat Development on Dredged Material," Technical Report DS-78-19, U.S. Army Engineer Waterways Experiment Station.

Standard Methods for the Examination of Water and Wastewater, 16th ed. (Washington, DC: American Public Health Association, 1985).

Stephenson, M. G. Turner, P. Pope, J. Colt, A. Knight and G. Tchobanoglous. "The Environmental Requirements of Aquatic Plants," (Sacremento: Calif. State Water Res. Control Bd. Appendix A of Pub. No. 65, Use and Potential of Aquatic Species for Wastewater Treatment, 1980).

Stuber, P. J., Coord. Proceedings of the National Symposium on Protection of Wetlands from Agricultural Impacts (Washington, D.C.: USDI, Fish and Wildlife Service Biol. Rep. 88(16), 1988).

Terrell, C. R., and P. B. Perfetti. "Water Quality Indicators Guide: Surface Waters," USDA, SCS; SCS-TP-161 (1989).

Teskey, R. O., and T. M. Hinckley. "Impact of Water Level Changes on Woody Riparian and Wetland Communities", USDI, Fish and Wildlife Service, Vol. I-VI; FWS/OBS-77/58 to 78/89 (1978).

Tiner, R. W., Jr. *Field Guide to Nontidal Wetland Identification* USDI, Fish and Wildlife Service (1988).

Veilleux, N., Ed. *National Wetlands Newsletter* (Washington, D.C.: Environmental Law Institute, various).

Weller, M. W. *Freshwater Marshes — Ecology and Wildlife Management,* 2nd ed. (Minneapolis: University of Minnesota Press, 1987).

Weller, M. W., Ed. *Waterfowl in Winter* (Minneapolis, University of Minnesota Press, 1988).

Wentz, W. A., R. L. Smith and J. A. Kadlec. "A Selected Annotated Bibliography on Aquatic and Marsh Plants and Their Management," (Ann Arbor, MI: University of Mich. School of Natural Resources, 1974).

Wetlands Services: The Ecology of . . . (Washington, D.C.: USDI, Fish and Wildlife Service, various) a comprehensive series on community profiles of important wetlands of the U.S. issued as FWS/OBS — xx.

Wetmacott, R. "Landscape and Wildlife Habitat Management in the Countryside," USDA, SCS; Landscape Architecture Note 3 (1987).

Wetzel, R. G. *Limnology* (Philadelphia: Saunders College Publishing, 1985).

Whitlow, T. H., and R. W. Harris. "Flood Tolerance in Plants: A State-of-the-Art Review," (Vicksburg, MS: U.S. Army Engineer Waterways Experiment Station, Technical Report E-79-2 (1979).

Wilcox, D. A., Ed. *Wetlands — The Journal of the Society of Wetland Scientists.* (Wilmington, NC: Soc. of Wetlands Sci.).

Wolf, R. B., L. C. Lee and R. R. Sharitz. "Wetland Creation and Restoration in the United States from 1970 to 1985: An Annotated Bibliography," (Wilmington, N.C.: The Society of Wetlands Scientists) *Wetlands* 6:1 (1986).

Zelanzny, J., and J. S. Feierabend. *Proceedings of a Conference: Increasing Our Wetland Resources* (Washington, D.C.: National Wildlife Federation, 1988).

APPENDIX C

COMMON AND SCIENTIFIC NAMES

PLANTS

Alder, Black	*Alnus glutinosa*
Alder, Hazel	*Alnus rugosa*
American Beech	*Fagus grandifolia*
Arrow Arum	*Peltandra cordata*
Arrowhead	*Sagittaria* spp.
Aspen, bigtooth	*Populus grandidentata*
Aspen, Quaking	*Populus tremuloides*
Baldcypress	*Taxodium distichum*
Barnyard Grass	*Echinochloa crusgalli*
Basswood	*Tilia americana*
Beggarticks	*Bidens* spp.
Birch, Paper	*Betula papyrifera*
Birch, White	*Betula populifolia*
Birch, Yellow	*Betula alleghaniensis*
Blackberries	*Rubus* spp.
Black Cherry	*Prunus serotina*
Black Gum	*Nyssa sylvatica*
Black Walnut	*Juglans nigra*
Bladderwort	*Utricularia* spp.
Bog Laurel	*Kalmia polifolia*
Box Elder	*Acer negundo*
Blueberry	*Vaccinium uliginosum*
Buck Bean	*Menyanthes trifoliata*
Bulrush, River	*Scirpus fluviatilis*
Bulrush, soft-stemmed	*Scirpus validus (S. actus)*
Burreed	*Sparganium eurycarpum*
Buttonbush	*Cephalanthus occidentalis*
Cattail, Narrow-leaved	*Typha angustifolia*
Cattail, Wide-leaved	*Typha latifolia*
China berry	*Melia azedarach*
Coontail	*Ceratophyllum demersum*
Cottonwood, Eastern	*Populus deltoides*

Cottonwood, Swamp	*Populus heterophylla*
Cranberry	*Vaccinium macrocarpon*
Dahoon Holly	*Ilex cassine*
Dog Fennel	*Eupatorium capilifolium*
Dogwood, Flowering	*Cornus florida*
Dogwood, Red-osier	*Cornus stolonifera*
Duckweed	*Lemna* spp.
Duckweed, Giant	*Spirodela* spp.
Elder	*Sambucus callicarpa*
Elm, American	*Ulmus americana*
Elm, Winged	*Ulmus alata*
Fern, Cinnamon	*Osmunda cinnamomea*
Fern, Marsh	*Thelypteris palustris*
Fern, Royal	*Osmunda regalis*
Fern, Water	*Azolla* spp.
Fetterbush	*Lyonia lucida*
Foxtail	*Alopecurus arundinaceus*
Foxtail Barley	*Hordeum jubatum*
Fragrant White Lily	*Nymphaea odorata*
Frogbit	*Limnobium spongia*
Gallberry	*Ilex glabra*
Giant Reed	*Phragmites australis*
Grass, Cotton	*Eriophorum polystachion*
Grass, Knot	*Paspalum* spp.
Grass, Marsh	*Scolochloa festucacea*
Grass, Manna	*Glyceria* spp.
Grass, Panic	*Panicum agrostoides*
Grass, Prairie Cord-	*Spartina pectinata*
Grass, Reed	*Calamogrostis inexpansa*
Grass, Reed Canary-	*Phalaris arundinacea*
Grass, Salt-	*Distichlis spicata*
Grass Saw-	*Cladium jamaicensis*
Grass, Slough-	*Beckmania syzigachne*
Grass, Switch-	*Panicum virgatum*
Grass, Wool-	*Scirpus cyperinus*
Grape	*Vitis* spp.
Green Ash	*Fraxinus pennsylvanica*
Greenbrier	*Smilax* spp.
Hackberry	*Celtis occidentalis*
Hardhack	*Spirea douglassii*
Hawthorn	*Crataegus mollis*
Hemlock, Eastern	*Tsuga canadensis*
Hemlock, Western	*Tsuga heterophylla*
Hemp	*Sesbania* spp.
Hempvine	*Mikania* spp.
Hickory, Bitternut	*Carya cordiformis*

Hiickory, Mockernut	*Carya tomentosa*
Hickory, Shagbark	*Carya ovata*
Hickory, Shellbark	*Carya lacinosa*
Hickory, Water	*Carya aquatica*
Holly, American	*Ilex opaca*
Holly, Deciduous	*Ilex decidua*
Honeysuckle	*Lonicera* spp.
Iris, Blue	*Iris virginicum*
Iris, Red	*Iris fulva*
Iris, Yellow	*Iris pseudacorus*
Ironwood	*Carpinus caroliniana*
Kentucky Coffee Tree	*Gymnocladus dioica*
Labrador Tea	*Ledum groenlandicum*
Lady's Slipper	*Cypripedium* spp.
Lizardtail	*Saururus cernuus*
Loblolly Bay	*Gordonia lasianthus*
Locust, Black	*Robinia pseudocacia*
Locust, Honey	*Gleditsia triacanthos*
Loosestrife	*Lysimachia* spp.
Maidencane	*Panicum hemitomon*
Mallow, Swamp	*Hibiscus moscheutos*
Mallow, Halbeard-leaved	*Hibiscus militaris*
Maple, Red	*Acer rubrum*
Maple, Silver	*Acer saccharinum*
Maple, Sugar	*Acer saccharum*
Marsh Marigold	*Caltha leptosepala*
Milfoil	*Myriophyllum* spp.
Milkweed, Swamp	*Asclepias incarnata*
Morning-glory	*Convolvulus* spp.
Moss, Sphagnum	*Sphagnum* spp.
Oak, Black	*Quercus velutina*
Oak, Bur	*Quercus macrocarpa*
Oak, Cherrybark	*Quercus falcata var. pagodifolia*
Oak, Laurel	*Quercus laurifolia*
Oak, Live	*Quercus virginiana*
Oak, Nuttal's	*Quercus nuttalii*
Oak, Overcup	*Quercus lyrata*
Oak, Pin	*Quercus palustris*
Oak, Red	*Quercus rubra*
Oak, Shingle	*Quercus imbricaria*
Oak, Shumard	*Quercus shumardii*
Oak, Spanish	*Quercus falcata*
Oak, Swamp White	*Quercus bicolor*
Oak, Valley	*Quercus lobata*
Oak, Water	*Quercus nigra*
Oak, White	*Quercus alba*

Oak, Willow	*Quercus phellos*
Orchid, Swamp	*Habenaria* spp.
Pecan	*Carya illinoensis*
Persimmon	*Diospyros virginiana*
Pickerelweed	*Pontederia cordata*
Pine, Pond	*Pinus serotina*
Pine, Slash	*Pinus elliottii*
Pine, Spruce	*Pinus glabra*
Pitcherplant	*Sarracenia* spp.
Poison Ivy	*Rhus radicans*
Pondweed, Sago	*Potamogeton pectinatus*
Pondweed, Sago	*Potamogeton pectinatus*
Popular, Fremont	*Populus fremontii*
Primrose, Willow	*Ludwidgia peruviana*
Quillwort	*Isoetes* spp.
Ragweed	*Ambrosia* spp.
Redbay	*Persea borbonia*
Rice Cutgrass	*Leersia oryzoides*
River Birch	*Betula nigra*
Sassafras	*Sassafras albidum*
Sedge	*Carex* spp.
Sedge, Chufa	*Cyperus* spp.
Sedge, Three-way	*Dulichium arundinaceum*
Skunk Cabbage	*Symplocarpus foetidus*
Skunk Cabbage, Yellow	*Lysichitum americanum*
Smartweed, Pale	*Polygonum lapathifolium*
Smartweed, Pennsylvania	*Polygonum pensylvanicum*
Smartweed, Swamp	*Polygonum coccineum*
Smartweed, Water	*Polygonum amphibium*
Soft Rush	*Juncus effusus*
Spatterdock	*Nuphar luteum*
Spider Lily	*Hymenocallis* spp.
Spikerush	*Eleocharis* spp.
Spruce, Black	*Picea mariana*
Spruce, Red	*Picea rubens*
Spruce, Sitka	*Picea sitchensis*
Sumac, Poison	*Toxicodendron vernix*
Sumac, Smooth	*Rhus glabra*
Swamp Privet	*Forestiera acuminata*
Swamp Rose	*Rosa palustris*
Sweetbay	*Magnolia virginiana*
Sweetflag	*Acorus calamus*
Sweetgum	*Liquidambar styraciflua*
Sugarberry	*Celtis laevigata*
Sycamore	*Platanus occidentalis*
Tamarack	*Larix laricina*

Tapegrass	*Vallisneria americana*
Three-square	*Scirpus americanus*
Trumpet Vine	*Campsis radicans*
Tupelo, Swamp	*Nyssa biflora*
Tupelo, Water	*Nyssa aquatica*
Virginia Creeper	*Parthenocissus quinquefolia*
Water Arum	*Calla palustris*
Watercress	*Nasturtium officinale*
Water Buttercup, Yellow	*Ranunculus flabellaris*
Water Buttercup, White	*Ranunculus aquatilis*
Water Elm	*Planera aquatica*
Water Heart	*Nymphoides aquatica*
Water Hyacinth	*Eichhornia crassipes*
Water Locust	*Gleditsia aquatica*
Water Lotus	*Nelumbo lutea*
Watermeal	*Wolffia* spp.
Water Pennywort	*Hydrocotyle umbellata*
Water Plantain	*Alisma* spp.
Water Primrose	*Jussiaea repens*
Water Shield	*Brasenia schreberi*
Water Starwort	*Callitriche* spp.
Water Weed	*Elodea* spp.
Waterwillow	*Justicia americana*
Wax Myrtle	*Myrica cerifera*
White Ash	*Fraxinus americana*
White Cedar, Atlantic	*Chamaecyparis thyoides*
White Cedar, Northern	*Thuja occidentalis*
Widgeongrass	*Ruppia maritima*
Wild Grape	*Vitis riparia*
Wild Rice	*Zizania aquatica*
Willow, Black	*Salix nigra*
Willow, Dune	*Salix piperi*
Willow, Hooker	*Salix hookeriana*
Willow, Narrow-leaf	*Salix exigua*
Willow, Pacific	*Salix lasiandra*
Wolffiella	*Wolffiella* spp.

ANIMALS

Bass	*Micropterus* spp.
Bluegill	*Lepomis macrochirus*
Bowfin	*Amia calva*
Bullhead	*Ictalurus* spp.
Carp	*Cyperinus carpio*
Catfish	*Ictalurus* spp.

Crappie	*Pomoxis* spp.
Fathead Minnow	*Pimephales promelas*
Gar *Lepisosteus* spp.	
Killifish	*Fundulus diaphanus*
Mosquitofish	*Gambusia affinis*
Northern pike	*Esox lucius*
Pickerel	*Esox americanus*
Shad	*Dorosoma cepedianum*
Shiner	*Notropis* spp.
Sunfish	*Lepomis cyanellus*
Sucker	*Catostomus* spp.
Top Minnow	*Pimephales* spp.
Walleye	*Stizostedion vitreum*
Yellow Perch	*Perca flavescens*
Bullfrog	*Rana catesbeiana*
Green Frog	*Rana clamitans*
Mudpuppy	*Necturus maculosus*
Tiger Salamander	*Ambystoma tigrinum*
Tree Frog	*Hyla* spp.
Water Siren	*Siren* spp.
Alligator	*Alligator mississipiensis*
Caiman	*Caiman* spp.
Snake, Garter	*Thamnophis* spp.
Snake, Mud	*Farancia abacura*
Snake, Queen	*Natrix septemvittata*
Snake, Water	*Natrix* spp.
Water Moccassin	*Ancistrodon piscivorus*
Turtle, Box	*Terrapene* spp.
Turtle, Cooter	*Pseudemys* spp.
Turtle, Mud	*Kinosternon* spp.
Turtle, Mask	*Sternotherus* spp.
Turtle, Painted	*Chrysemys* spp.
Turtle, Pond	*Clemmys* spp.
Turtle, Slider	*Pseudemys* spp.
Turtle, Snapping	*Chelydra serpentina*
Turtle, Softshell	*Trionyx* spp.
American Coot	*Fulica americana*
Bittern, American	*Botaurus lentiginosus*
Blackbird	*Xanthocephalus* spp.
Canada goose	*Branta canadensis*
Chickadee	*Parus* spp.
Cormorant	*Phalacrocorax* spp.
Crane	*Gruidae*

Duck, Dabbling	*Anatids*
Duck, Diving	*Aythyatids*
Duck, Wood	*Aix sponsa*
Eagle, Bald	*Haliaeetus leucocephalus*
Flycatcher	*Tyrannidae*
Gadwall	*Anas strepera*
Grebe	*Podiceps* spp.
Gull	*Larus* spp.
Heron	*Ardea* spp.
Loon	*Gavia* spp.
Mallard	*Anas platyrhynchos*
Osprey	*Pandion haliaetus*
Pelican, Brown	*Pelecanus occidentalis*
Pelican, White	*Pelecanus erythrorhynchos*
Peregrine Falcon	*Falco peregrinus*
Pheasant	*Phasianus colchicus*
Rail	*Rallidae*
Red-headed Woodpecker	*Melanerpes erythrocephalus*
Shorebird	*Charadrii*
Short-eared Owl	*Asio flammeus*
Snail Kite	*Rostrhamus sociabilis*
Sparrow, Swamp	*Melospiza georgiana*
Swan, Mute	*Cygnus olor*
Swan, Trumpeter	*Olor buccinator*
Warbler	*Parulidae*
Widgeon	*Mareca americana*
Wren	*Cistothorus* spp.
Bear, Black	*Ursus americanus*
Beaver	*Castor canadensis*
Bobcat	*Lynx rufus*
Coyote	*Canis latrans*
Deer, White-tailed	*Odocoileus virginianus*
Deer, Mule	*Odocoileus hemionus*
Elk	*Cervus canadensis*
Lemming	*Synaptomys* spp.
Marsh Rice Rat	*Oryzomys palustris*
Mink	*Mustela vision*
Muskrat	*Ondatra zibethica*
Moose	*Alces alces*
Nutria	*Myocaster coypus*
Opossum	*Didelphis marsupialis*
Otter	*Lutra canadensis*
Pig, Feral	*Sus scrofa*
Raccoon	*Procyon lotor*
Skunk	*Mephitis* or *Spilogale* spp.

Swamp Rabbit	*Sylvilagus aquaticus*
Water Shrew	*Sorex palustris*
Weasel	*Mustela* spp.
Wolf	*Canis lupus*
Wood Rat	*Neotoma* spp.

INDEX